Hungarian Problem Book III

Hungarian Problem Book III

Based on the Eötvös Competition 1929–1943
Compiled by G. Hájos, G. Neukomm, and J. Surányi

Translated and edited by

Andy Liu

Published and Distributed by
The Mathematical Association of America

The New Mathematical Library (NML) was started in 1961 by the School Mathematics Study Group to make available to high school students short expository books on various topics not usually covered in the high school syllabus. In a decade the NML matured into a steadily growing series of some twenty titles of interest not only to the originally intended audience, but to college students and teachers at all levels. Previously published by Random House and L. W. Singer, the NML became a publication series of the Mathematical Association of America (MAA) in 1975. Under the auspices of the MAA the NML continues to grow and remains dedicated to its original and expanded purposes. In its third decade, it contains forty titles.

ANNELI LAX NEW MATHEMATICAL LIBRARY

Books may be ordered from:
MAA Service Center
P. O. Box 91112
Washington, DC 20090-1112
1-800-331-1622   fax: 301-206-9789

Foreword

It's a pleasure for me to write a Foreword to this *Hungarian Problem Book III* which covers the Eötvös/Kürschák competitions of the period 1929–1943. As you have all the mathematics of it in the book itself, I would like to say a few words first on the present organization of the competition and then on the history of it.

The competition is organized every year on an October afternoon and every high-school student can take part in it, including students who have just finished high school that summer. It is a four-hour contest with three problems to solve. Students can use anything they bring with them such as books and notes. However use of computers and calculators is not permitted. The competition takes place simultaneously in 20 cities all over Hungary.

I cite the key sentence of the "Call for Participation" of the competition: "According to the traditions of the competition, the solution of the competition problems requires no knowledge beyond the usual high-school material; however, finding the solution challenges the student's ability for independent mathematical thinking."

Of course, before and after the competition, there is a lot of work to be done. There is a Competition Committee (henceforth abbreviated to CC) consisting of about 10 people. (The fact that I am a member of this CC largely facilitates my writing this account.)

The work starts in the summer by collecting proposed problems. These are mostly created by members of the CC themselves, but quite often research mathematicians in Hungary offer problem proposals to the CC. (I know from discussion with foreign colleagues that many countries are not so lucky in this respect.) Previous years' unused problems are also included in the set of proposed problems (they are always kept secret for future use).

At the end of the process a collection of about 10 to 20 problems emerges. However, usually not all of them are ready for use: members of the CC who have sons or daughters in high school that year (this is quite a common phenomenon) invariably ask that the problems proposed by them not be used until their offspring are ineligible.

Roughly one month before the actual competition, the CC meets. I would consider ourselves very lucky if we can choose the three problems after just one lengthy discussion. Otherwise, a second meeting must be convened. After having agreed on the three contest problems, the CC determines how to formulate them, but this is usually only a matter of minutes.

After the competition has taken place, the other 19 cities send the papers written there to Budapest where all the contest papers are distributed into equal packages and each member of the CC receives one such package. He eliminates the papers having no chance for a prize and circulates those which have a chance. (Papers with two complete solutions are almost invariably circulated.) On rare occasions, a paper which otherwise has no chance, but contains a nice or unusual solution is also circulated. Such papers do not get a prize, but the solutions may eventually be presented at the prize-giving ceremony.

When all the circulated papers are collected, they are taken home for two or three days by every member of the CC for careful inspection. So when the final meeting of the CC convenes, it is usually an easy matter, since everybody is in full knowledge of the details.

At the beginning of the competition (in 1894) and for many years afterwards, there was one first and one second prize. However in those years the number of participants was around 30. In the '50s and after, the number of participants increased to several hundreds. It became necessary to allow more than one first prize and more than one second prize, and in some years even third prizes were given. Also in some years there were honourable mentions. Of course it is still possible to have just one first prize and it may even happen (very rarely) that no first prize at all is given. The CC has great flexibility in this respect; we just try to find the best way to express the outcome of the competition.

There are small cash prizes given to the winners; they are provided by the János Bolyai Mathematical Society. (There is no entry fee for the competition.) The whole procedure ends with a prize-giving ceremony, where the Chairman of the CC presents the solutions of the problems (quite often more than one solution to the same problem).

Having described the present state of the competition, let me say a few words about its history.

First, about the curious Eötvös/Kürschák duality of its name, which in my experience confuses many people. In the year 1894, when the competition started, there was a Mathematical *and* Physical Society in Hungary and Loránd Eötvös (himself a physicist) was the chairman. When he was appointed Minister of Education that year, the Society decided to celebrate the occasion by organizing a competition for high-school students which then became a tradition. The competition had that name till 1943, when war interrupted it. After the war the Mathematical and Physical Societies split and the physicists started to organize their own competition. It was only natural that they had the right to the name Eötvös, so for the mathematics competition a new name had to be found. A natural choice was József Kürschák, a professor of mathematics at the Technical University of Budapest, a keen supporter of mathematics competitions and author of — among other books for high-school students — the book which was translated as *Hungarian Problem Books I* and *II*.

In the period covered by this book (and before this period, too), the number of participating students rarely exceeded 30. (Incidentally in those years high-school students were not admitted; only students who had just graduated from high school were eligible.) However, after the war the number of participants increased drastically to several hundreds, sometimes coming close to one thousand. The reasons for this phenomenon are too complex to be analyzed here, so I mention just one of them, which may not come immediately to mind.

In those years, every university in Hungary had a strict entrance examination. However, if you were in the first ten of the Kürschák competition, then you had free access to any university where mathematics was one of the entrance examination subjects. This certainly was very attractive to those students who felt they had a chance to gain entrance this way.

After 1990 the entrance examination system was largely abolished (now it varies from university to university and even within a single university). So this attraction disappeared. Also, with the gradual introduction of a market economy in Hungary since 1990, professions like banking, law and economics gained in popularity, and correspondingly interest in mathematics declined. These days only those students who are genuinely interested in mathematics attempt the Kürschák competition. The number of participants has varied between 100 and 300 in recent years.

Nevertheless, it is still true that the best students — those whom you will hear about 10 or 20 years from now — invariably take part in the Kürschák competition. To illustrate that there has always been a strong correlation between Kürschák prize winners and mathematicians attaining fame, let me give you the names of some prize winners in the 1929–1943 period who later became internationally known mathematicians: Tibor Gallai (1930), Endre Makai (1933), Tibor Szele (1936) and Ákos Császár (1942). As I said, in those years, the number of participants was small. When the continuation to this book appears, you will be astonished at the appearance of the names of several dozen well-known (sometimes world-famous) mathematicians among the prize-winners.

I think English-speaking people who have an interest in mathematics (not just competitions) can only be grateful to Andy Liu for translating this book (and adding notes, making it accessible to even more people). I am also grateful to him for his translation, but for different reasons; I read the Hungarian original several decades ago. In conclusion, I can only say: I hope that *Hungarian Problem Book IV* will soon follow.

József Pelikán,
Budapest, 2000.

Contents

Preface

The Eötvös Mathematics Competition is the oldest in the world for high school students organized on a national scale, with a tradition dating back to 1894. Four volumes have appeared in Hungarian, covering the periods 1894–1928, 1929–1963, 1964–1987 and 1988–1997. In 1949, the contest was renamed the Kürschák Mathematics Competition.

In 1963, the New Mathematical Library published the translation of the first Hungarian volume in two books, titled *The Hungarian Problem Book I* and *Book II*. They are #11 and #12 in the series respectively. However, no further translations have appeared since then, even after the Mathematical Association of America took over the publication of the series. The delay is a long and involved story not worth recording.

In 1995, I was approached by Underwood Dudley, then editor of the series, to work on *Hungarian Problem Book III*. I was both excited and apprehensive. My very limited knowledge of the Hungarian language was certainly not sufficient for the task. Fortunately, I had acquired over the years various documents in English and Chinese pertaining to this contest. János Csirik, a Hungarian graduate student in mathematics at the University of California, Berkeley, was on call via email whenever I had translation difficulties.

The competition was not held in 1944–1946 because of the Second World War. The present book takes the contest into 1943, about midway through the second Hungarian volume. I am planning to complete the translation of that volume in a subsequent book.

There are primarily two types of books on problem-solving. One consists of books on the techniques, with illustrative examples from various sources. Experienced students often complain that they repeat well-known problems.

The other consists of problem collections — they are often tied to a specific contest, but there are also many problem books with less well-defined domains. Beginners find it hard to learn problem-solving from them.

I have always wanted to do a book that can serve both beginners and experienced students. *The Hungarian Problem Book III* seems an ideal opportunity. I want to make it clear that this is not out of any disrespect to the original work.

I visited Professor János Surányi, principal author of the four Hungarian volumes, in the spring of 1999. I showed him a draft of my work, and was gratified that he approved of what I had attempted.

In the original work, multiple solutions were often offered. I have kept this feature. Sometimes, solutions were followed by discussions of varying length, either going over necessary background material, providing generalizations or making relevant remarks about the problems. I have moved much of this material to have it appear before the solutions.

The present book consists of six chapters. In the first, the contest problems are given in chronological order. There are 45 problems altogether. They are classified by subject into five groups: Combinatorics, Number Theory, Algebra, Geometry – Part I and Geometry – Part II. Chapter 5 contains geometry of the polygon, while Chapter 6 contains geometry of the circle, as well as other topics related to geometry.

Experienced students can simply attempt the problems and then compare their work with the solutions offered. A Problem Index with page references for the solutions of individual problems appears on page xvii and page 142.

Beginners can use the present book in a different way. They are encouraged to cover Chapters 2 to 6 in order. In Chapter 2, the nine Combinatorics problems are restated and it is suggested that the students make an initial attempt to solve them in the given order. If they achieve some success, they can check their work against the solutions presented in Section 2.2.

On the other hand, if they find the problems hard to tackle, they can read Section 2.1 which gives detailed discussions. Basic definitions, important results and various problem-solving techniques are presented. Chapters 3 to 6 are organized in the same fashion. The last two chapters are particularly detailed, since the presence of geometry in the North American high school curriculum has significantly diminished. Much of the material here is based on the textbook *Euclidean Geometry* by Ed Leonard, Ted Lewis, Andy Liu and George Tokarsky.

It should be pointed out that the discussion sections are not meant to be read in one sitting. They are primarily provided as references. Teachers who may wish to base a training program on these sections should provide additional illustrations and exercises.

As mentioned earlier, I have incorporated much of the discussions in the original volume, but I have also left out some of the more advanced material that is not needed for tackling the problems at hand. Instead, I added some carefully chosen elementary material. I tried to make sure that everything in the discussion sections is indeed relevant to at least one of the actual contest problems.

There are also a few solutions that were not in the original work. These were provided by two independent groups. The first consisted of nine high school students in Edmonton, under the leadership of Byung-Kyu Chun. The other members were Kathy Wong, Vivian Yu, Gilbert Lee, Robert Lutz, Philip Stein, Frank Chen, Anton Cherney and Bernie Kan. All have since entered universities.

The second group consisted of two members of the People's Republic of China's team in the 1999 International Mathematical Olympiad, Ruochuan Liu and Xin Li. Both won gold medals in helping China tie with Russia for first place in the unofficial team standing. Vardges Levonian, an Armenian undergraduate student in mathematics at Brigham Young University, read various parts of the manuscript and suggested improvements.

Underwood Dudley gave me encouragement and support until he relinquished his editorship to Michael McAsey, who continued in the same positive role. Anneli Lax provided a very thorough and critical review of Chapter 2 before her untimely illness. The daunting task of combing through the remaining part of the manuscript fell into the lap of Richard Guy and George Berzsenyi, with valuable input from other members of the editorial committee. In particular, Richard Guy provided tremendous help during a three-day session in Calgary, mathematically, linguistically, and technically. I would like to thank Elaine Pedreira and Beverly Ruedi of the Mathematical Association of America for editorial and technical support.

Murray Klamkin, who needs no introduction in the world of problem-solving, was consulted on a regular basis, both for his incisive insight and encyclopedic knowledge. Finally, I want to thank my Hungarian friend József Pelikán for writing the Foreword. He also provided invaluable help on various matters at different stages of this project. I would also like to acknowledge the unidentified Hungarian mathematicians who proposed such inspiring problems in the first place.

Andy Liu
Edmonton, 2000

Problem Index

Year	#1		#2		#3	
1929	CB	27	AB	61	G1	103
1930	CB	26	CB	25	G1	93
1931	NT	43	NT	44	G1	91
1932	NT	46	G2	127	G2	140
1933	AB	56	CB	24	G2	128
1934	NT	46	G2	137	CB	31
1935	AB	59	CB	30	CB	29
1936	AB	56	G1	98	NT	47
1937	AB	59	G2	144	G1	92
1938	NT	43	AB	59	G2	142
1939	AB	61	NT	45	G2	130
1940	CB	23	NT	44	G1	97
1941	AB	57	G2	139	G2	132
1942	G1	91	NT	43	G1	99
1943	CB	24	G1	94	AB	58

We give each problem a two-character code for its subject classification, and a number which indicates the page where its solution begins.

Code	Subject	Chapter
AB	Algebra	4
CB	Combinatorics	2
G1	Geometry Part 1	5
G2	Geometry Part 2	6
NT	Number Theory	3

This chart is repeated on page 142.

List of Winners

1929	**Párducz** Nándor,	**Szolovics** Dezső.
1930	**Gallai** Tibor,	**Démán** Pál.
1931	**Sebők** György,	**Ernst** Frigyes.
1932	**Alpár** László.	
1933	**Makai** Endre,	**Felter** Károly.
1934	**Vázsonyi** Endre.	
1935	**Bella** Andor,	**Csanádi** György.
1936	**Szele** Tibor,	**Kepes** Ádám.
1937	**Frankl** Ottó.	
1938	**Bodó** Zalán,	**Weisz** Alfréd.
1939	**Sándor** Gyula,	**Csáki** Frigyes.
1940	**Hoffmann** Tibor,	**Bizám** György.
1941	**Ivancsó** Imre,	**Schweitzer** Miklós.
1942	**Császár** Ákos,	**Pál** Sándor.
1943	**Ádám** István,	**Takács** Lajos.

1

Eötvös Mathematics Competition Problems

1929

Problem 1.

In how many ways can the sum of 100 fillér be made up with coins of denominations 1, 2, 10, 20 and 50 fillér?

Problem 2.

Let $k \leq n$ be positive integers and x be a real number with $0 \leq x < \frac{1}{n}$. Prove that

$$\binom{n}{0} - \binom{n}{1}x + \binom{n}{2}x^2 - \cdots + (-1)^k \binom{n}{k}x^k > 0.$$

Problem 3.

Let p, q and r be three concurrent lines in the plane such that the angle between any two of them is $60°$. Let a, b and c be real numbers such that $0 < a \leq b \leq c$.

(a) Prove that the set of points whose distances from p, q and r are respectively less than a, b and c consists of the interior of a hexagon if and only if $a + b > c$.

(b) Determine the length of the perimeter of this hexagon when $a + b > c$.

Remark: To locate solutions to the problems, refer to the Problem Index on page xvii or page 142.

1

1930

Problem 1.

How many five-digit multiples of 3 end with the digit 6?

Problem 2.

A straight line is drawn across an 8×8 chessboard. It is said to pierce a square if it passes through an interior point of the square. At most how many of the 64 squares can this line pierce?

Problem 3.

Inside an acute triangle ABC is a point P that is not the circumcenter. Prove that among the segments AP, BP and CP, at least one is longer and at least one is shorter than the circumradius of ABC.

1931

Problem 1.

Let p be a prime greater than 2. Prove that $\frac{2}{p}$ can be expressed in exactly one way in the form $\frac{1}{x} + \frac{1}{y}$ where x and y are positive integers with $x > y$.

Problem 2.

Let $a_1^2 + a_2^2 + a_3^2 + a_4^2 + a_5^2 = b^2$, where a_1, a_2, a_3, a_4, a_5 and b are integers. Prove that not all of these numbers can be odd.

Problem 3.

Let A and B be two given points, distance 1 apart. Determine a point P on the line AB such that $\frac{1}{1+AP} + \frac{1}{1+BP}$ is a maximum.

1932

Problem 1.

Let a, b and n be positive integers such that b is divisible by a^n. Prove that $(a+1)^b - 1$ is divisible by a^{n+1}.

Problem 2.

In triangle ABC, $AB \neq AC$. Let AF, AP and AT be the median, angle bisector and altitude from vertex A, with F, P and T on BC or its extension.

(a) Prove that P always lies between F and T.

(b) Prove that $\angle FAP < \angle PAT$ if ABC is an acute triangle.

Problem 3.

Let α, β and γ be the interior angles of an acute triangle. Prove that if $\alpha < \beta < \gamma$, then $\sin 2\alpha > \sin 2\beta > \sin 2\gamma$.

1933

Problem 1.

Let a, b, c and d be real numbers such that $a^2 + b^2 = c^2 + d^2 = 1$ and $ac + bd = 0$. Determine the value of $ab + cd$.

Problem 2.

Sixteen squares of an 8×8 chessboard are chosen so that there are exactly two in each row and two in each column. Prove that eight white pawns and eight black pawns can be placed on these sixteen squares so that there is one white pawn and one black pawn in each row and in each column.

Problem 3.

The circles k_1 and k_2 are tangent at the point P. A line is drawn through P, cutting k_1 at A_1 and k_2 at A_2. A second line is drawn through P, cutting k_1 at B_1 and k_2 at B_2. Prove that the triangles PA_1B_1 and PA_2B_2 are similar.

1934

Problem 1.

Let n be a given positive integer and

$$A = \frac{1 \cdot 3 \cdot 5 \cdots (2n-1)}{2 \cdot 4 \cdot 6 \cdots 2n}.$$

Prove that at least one term of the sequence $A, 2A, 4A, 8A, \ldots, 2^k A, \ldots$ is an integer.

Problem 2.

Which polygon inscribed in a given circle has the property that the sum of the squares of the lengths of its sides is maximum?

Problem 3.

We are given an infinite set of rectangles in the plane, each with vertices of the form $(0,0)$, $(0,m)$, $(n,0)$ and (n,m), where m and n are positive integers. Prove that there exist two rectangles in the set such that one contains the other.

1935

Problem 1.

Let n be a positive integer. Prove that

$$\frac{a_1}{b_1} + \frac{a_2}{b_2} + \cdots + \frac{a_n}{b_n} \geq n,$$

where $\langle b_1, b_2, \ldots, b_n \rangle$ is any permutation of the positive real numbers $a_1, a_2, \ldots, a_n$.

Problem 2.

Prove that a finite point set cannot have more than one center of symmetry.

Problem 3.

A real number is assigned to each vertex of a triangular prism so that the number on any vertex is the arithmetic mean of the numbers on the three adjacent vertices. Prove that all six numbers are equal.

1936

Problem 1.

Prove that for all positive integers n,

$$\frac{1}{1 \cdot 2} + \frac{1}{3 \cdot 4} + \cdots + \frac{1}{(2n-1)2n} = \frac{1}{n+1} + \frac{1}{n+2} + \cdots + \frac{1}{2n}.$$

Problem 2.

S is a point inside triangle ABC such that the areas of the triangles ABS, BCS and CAS are all equal. Prove that S is the centroid of ABC.

Problem 3.

Let a be any positive integer. Prove that there exists a unique pair of positive integers x and y such that $x + \frac{1}{2}(x + y - 1)(x + y - 2) = a$.

1937

Problem 1.

Let n be a positive integer. Prove that $a_1! \, a_2! \cdots a_n! < k!$, where k is an integer which is greater than the sum of the positive integers $a_1, a_2, \ldots, a_n$.

Problem 2.

Two circles in space are said to be tangent to each other if they have a common tangent at the same point of tangency. Assume that there are three circles in space which are mutually tangent at three distinct points. Prove that they either all lie in a plane or all lie on a sphere.

Problem 3.

Let n be a positive integer. Let P, Q, A_1, $A_2, \ldots, A_n$ be distinct points such that A_1, $A_2, \ldots, A_n$ are not collinear. Suppose that $PA_1 + PA_2 + \cdots + PA_n$ and $QA_1 + QA_2 + \cdots + QA_n$ have a common value s for some real number s. Prove that there exists a point R such that

$$RA_1 + RA_2 + \cdots + RA_n < s.$$

1938

Problem 1.

Prove that an integer n can be expressed as the sum of two squares if and only if $2n$ can be expressed as the sum of two squares.

Problem 2.

Prove that for all integers $n > 1$, $\dfrac{1}{n} + \dfrac{1}{n+1} + \cdots + \dfrac{1}{n^2 - 1} + \dfrac{1}{n^2} > 1$.

Problem 3.

Prove that for any acute triangle, there is a point in space such that every line segment from a vertex of the triangle to a point on the line joining the other two vertices subtends a right angle at this point.

1939

Problem 1.

Let a_1, a_2, b_1, b_2, c_1 and c_2 be real numbers for which $a_1 a_2 > 0$, $a_1 c_1 \geq b_1^2$ and $a_2 c_2 \geq b_2^2$. Prove that $(a_1 + a_2)(c_1 + c_2) \geq (b_1 + b_2)^2$.

Problem 2.

Determine the highest power of 2 that divides $2^n!$.

Problem 3.

ABC is an acute triangle. Three semicircles are constructed outwardly on the sides BC, CA and AB respectively. Construct points A', B' and C' on these semicircles respectively so that $AB' = AC'$, $BC' = BA'$ and $CA' = CB'$.

1940

Problem 1.

In a set of objects, each has one of two colors and one of two shapes. There is at least one object of each color and at least one object of each shape. Prove that there exist two objects in the set that are different both in color and in shape.

Problem 2.

Let m and n be distinct positive integers. Prove that $2^{2^m} + 1$ and $2^{2^n} + 1$ have no common divisor greater than 1.

Problem 3.

(a) Prove that for any triangle H_1, there exists a triangle H_2 whose side lengths are equal to the lengths of the medians of H_1.

(b) If H_3 is the triangle whose side lengths are equal to the lengths of the medians of H_2, prove that H_1 and H_3 are similar.

1941

Problem 1.

Prove that

$$(1+x)(1+x^2)(1+x^4)(1+x^8)\cdots(1+x^{2^{k-1}}) = 1+x+x^2+x^3+\cdots+x^{2^k-1}.$$

Problem 2.

Prove that if all four vertices of a parallelogram are lattice points and there are some other lattice points in or on the parallelogram, then its area exceeds 1.

Problem 3.

The hexagon $ABCDEF$ is inscribed in a circle. The sides AB, CD and EF are all equal in length to the radius. Prove that the midpoints of the other three sides determine an equilateral triangle.

1942

Problem 1.

Prove that in any triangle, at most one side can be shorter than the altitude from the opposite vertex.

Problem 2.

Let a, b, c and d be integers such that for all integers m and n, there exist integers x and y such that $ax + by = m$ and $cx + dy = n$. Prove that $ad - bc = \pm 1$.

Problem 3.

Let A', B' and C' be points on the sides BC, CA and AB, respectively, of an equilateral triangle ABC. If $AC' = 2C'B$, $BA' = 2A'C$ and $CB' = 2B'A$, prove that the lines AA', BB' and CC' enclose a triangle whose area is $\frac{1}{7}$ that of ABC.

1943

Problem 1.

Prove that in any group of people, the number of those who know an odd number of the others in the group is even. Assume that "knowing" is a symmetric relation.

Problem 2.

Let P be any point inside an acute triangle. Let D and d be respectively the maximum and minimum distances from P to any point on the perimeter of the triangle.

(a) Prove that $D \geq 2d$.

(b) Determine when equality holds.

Problem 3.

Let $a < b < c < d$ be real numbers and $\langle x, y, z, t \rangle$ be any permutation of a, b, c and d. What are the maximum and minimum values of the expression $(x - y)^2 + (y - z)^2 + (z - t)^2 + (t - x)^2$?

Remark: As we said at the beginning, to locate solutions to the problems, refer to the Problem Index on page xvii or page 142.

2
Combinatorics Problems

Problem 1940.1.

In a set of objects, each has one of two colors and one of two shapes. There is at least one object of each color and at least one object of each shape. Prove that there exist two objects in the set that are different both in color and in shape.

Problem 1943.1.

Prove that in any group of people, the number of those who know an odd number of the others in the group is even. Assume that "knowing" is a symmetric relation.

Problem 1933.2.

Sixteen squares of an 8×8 chessboard are chosen so that there are exactly two in each row and two in each column. Prove that eight white pawns and eight black pawns can be placed on these sixteen squares so that there is one white pawn and one black pawn in each row and in each column.

Problem 1930.2.

A straight line is drawn across an 8×8 chessboard. It is said to pierce a square if it passes through an interior point of the square. At most how many of the 64 squares can this line pierce?

Problem 1930.1.

How many five-digit multiples of 3 end with the digit 6?

Problem 1929.1.

In how many ways can the sum of 100 fillér be made up with coins of denominations 1, 2, 10, 20 and 50 fillér?

Problem 1935.3.

A real number is assigned to each vertex of a triangular prism so that the number on any vertex is the arithmetic mean of the numbers on the three adjacent vertices. Prove that all six numbers are equal.

Problem 1935.2.

Prove that a finite point set cannot have more than one center of symmetry.

Problem 1934.3.

We are given an infinite set of rectangles in the plane, each with vertices of the form $(0,0)$, $(0,m)$, $(n,0)$ and (n,m), where m and n are positive integers. Prove that there exist two rectangles in the set such that one contains the other.

2.1 Discussion

2.1.1 Problem-solving

Combinatorics problems usually do not require a lot of technical knowledge. This in some sense makes them more difficult, because we have to rely more on intuition and insight. Often, we do not even know where to begin on such problems.

For this, we turn to the advice of one of the greatest problem solvers. In his book *How To Solve It*, George Pólya outlined a four-step scheme:

1. Understand the problem.
2. Devise a plan.
3. Carry out the plan.
4. Look back.

On the surface, this scheme seems trivial. In the book, these four steps are elaborated further. However, they become truly alive when we apply them in solving actual problems. We will deal with steps 1 and 2 in this subsection, and steps 3 and 4 in the next.

Let us look through the nine problems and see if there is any difficulty in understanding them. The statement of **Problem 1940.1** is in everyday language and fairly straightforward. The same holds true for **Problem 1943.1**, except possibly the term *symmetric relation*.

A **binary relation** R on a set S is a collection of ordered pairs of elements of S. If a and b are elements of S that are not necessarily distinct, and the ordered pair (a, b) belongs to R, then we write aRb and say "a is related to b." The statement aRb is either true or false, according to whether (a, b) belongs to R or not.

Technically, any such collection defines a binary relation. More often, a rule is specified that determines whether a particular ordered pair belongs to the collection. For example, if S is the set of real numbers and R stands for *is equal to*, then $5 = 3$ is false and $4 = 4$ is true. If instead R stands for *is greater than or equal to*, then $5 \geq 3$ is true and $4 \geq 4$ is also true.

The binary relation R is said to be **symmetric** if aRb implies bRa. For example, "is equal to" is symmetric while "is greater than or equal to" is not. It is necessary to state explicitly that *"knowing" is a symmetric relation*, because in everyday language, it is possible to interpret the relationship of "knowing" in the following unintended way. We may claim to "know" certain celebrities, although it is virtually certain that they have never heard of us.

It is the task of the problem-poser to eliminate such ambiguities. However, it may happen in some contest that you find a problem that can be interpreted in at least two different ways. Usually, one of them leads to a trivial situation, and it is obviously not what the problem-poser has intended. You may also have read the problem only in that way. You should resist the temptation of getting a quick way out, and look for an alternative and reasonable interpretation. In any case, you should write down what you think the problem is asking and proceed to attack it accordingly.

In **Problem 1930.2**, an interior point of the square means a point of the square that is not on its perimeter. Thus the line must run through the square, and not just along a side or through a corner.

You should have no further difficulty until you come to the term *fillér* in **Problem 1929.1**. It is a unit of Hungarian currency and fillér is also used as the plural. However, you should realize that its precise meaning is irrelevant to the problem, which is unchanged if we replace it by *cent* or some nonsense word like *cthulhu*. Actually, the original statement asks in how many ways can change be given for 1 forint. It is then necessary to know that in Hungarian currency, 1 forint is equal to 100 fillér.

In **Problem 1935.3**, there is a technical term, **arithmetic mean**. It is synonymous with the ordinary *average*, and the arithmetic mean of the real numbers $x_1, x_2, \ldots, x_n$ is defined to be

$$M_1(x_1, x_2, \ldots, x_n) = \frac{x_1 + x_2 + \cdots + x_n}{n}.$$

The reason why we use this longer name is because there are other reasonable ways of measuring the average of these numbers. The notation M_1 is used to distinguish the arithmetic mean from the others.

All **mean functions** $M(x_1, x_2, \ldots, x_n)$ have the property that $M(x, x, \ldots, x) = x$, though this alone does not define a mean function. We now look at two other mean functions, which at first do not sound like reasonable ways of measuring the average. The largest of the numbers $x_1, x_2, \ldots, x_n$ is called the **maximum** and is denoted by $M_\infty(x_1, x_2, \ldots, x_n) = \max\{x_1, x_2, \ldots, x_n\}$. Observe that we do have $M_\infty(x, x, \ldots, x) = x$. The smallest of these n numbers is called the **minimum**, $M_{-\infty}(x_1, x_2, \ldots, x_n) = \min\{x_1, x_2, \ldots, x_n\}$. We also have $M_{-\infty}(x, x, \ldots, x) = x$.

We are now ready to state part of an important result called the **Power Means Inequality**. We will say more about it in Chapter 4, where more mean functions will be defined. The three that we have so far are in a specific order. More precisely, for any real numbers $x_1, x_2, \ldots, x_n$, we have

$$M_{-\infty}(x_1, x_2, \ldots, x_n) \leq M_1(x_1, x_2, \ldots, x_n) \leq M_\infty(x_1, x_2, \ldots, x_n).$$

Let us prove the second part of the inequality. Let x denote the maximum of these n numbers. Then $x_i \leq x$ for $1 \leq i \leq n$. Adding these inequalities yields $x_1 + x_2 + \cdots + x_n \leq nx$. Dividing both sides by the positive number n yields the desired result.

Note that we have used $\leq$ instead of $<$, thus keeping open the possibility that the arithmetic mean may be equal to the maximum. This can happen only if all n numbers are equal. Moreover, if the numbers are indeed identical, then the arithmetic mean will be equal to the maximum. We therefore say that equality holds if and only if $x_1 = x_2 = \cdots = x_n$.

It should be clear that the first part of the inequality can be proved in the same way. In a competition, you may simply say so if you are absolutely sure that this is the case. It will save time not to repeat the same argument all over again. In this instance, the only change is the replacement of "maximum" by "minimum," $\leq$ by $\geq$, and M_∞ by $M_{-\infty}$. However, if there are other minor but necessary modifications to an argument, they should be stated clearly.

After this digression, we note that there is one other technical term, in **Problem 1935.2**. A **center of symmetry** of a point set S is a point O, not necessarily in S, so that for any point A in S, there is also a point B in S where O is the midpoint of the line segment AB. We say that B is *symmetric* to A with respect to O.

Note that the dimension of the space in which the point set is situated is not actually specified; the definition is valid in any dimension. If the dimension is at least two, a useful result for tackling this problem is the **Triangle Inequality**, which states that the sum of the lengths of two sides of a triangle is at least as great as the length of the third side. However, if you can only solve the problem in one dimension, you should still write down your solution because you may get some partial credit.

Now that we have a clear understanding of the problems, we have to formulate means of attack. For some problems, this step may be obvious, even though seeing it through to the end may not be at all easy. For other problems, we may get stuck right here.

A useful general technique is to solve first a similar but simpler problem, and try to gather useful information from it. You may spot a pattern that may suggest a way of tackling the original problem. For instance, in **Problem 1935.3**, we replace the triangular prism by a triangle, thereby reducing the problem from three dimensions to two dimensions.

Let the numbers on the three vertices be a, b and c. The analogous condition is that each of them is the arithmetic mean of its two neighbors. Hence

$$2a = b + c,$$

$$2b = c + a,$$

$$2c = a + b.$$

Subtracting the second equation from the first, we have $2a - 2b = b - a$ or $3a = 3b$. Hence $a = b$. Substituting $a = b$ into the first equation, we have $2b = b + c$ so that $b = c$.

Why do we not need all three equations? This is because they are not *independent* of one another. If you add any two of them, you will end up with the third. The concept of independence is very important, and we will come across several instances of it.

Combinatorics is undergoing rapid development in the computer age. Bodies of knowledge have begun to take shape. In the following subsections, we present some of the basic results most useful in problem-solving.

2.1.2   Graph Theory

Sometimes, what appears to be a very difficult problem may become quite easy once you look at it in the right way. Thus the ability to express problems in different forms is very useful. This is known as *setting up mathematical models*.

A very versatile model is a **graph**. Defined formally, a graph consists of a set V of elements, called **vertices**, and a set E of pairs of vertices, called **edges**. The edge $\langle u, v \rangle$ is said to **join** the two vertices u and v. These two vertices are said to be **adjacent** and are called the **endpoints** of $\langle u, v \rangle$. Two edges are said to be **adjacent** if they have a common endpoint.

A set of vertices or of edges is said to be **independent** if no two in the set are adjacent. A **subgraph** of a graph consists of a subset V' of V and a subset E' of those edges which join two vertices of V'.

The vertices and edges can stand for virtually anything. For instance, consider a convex polyhedron. Its skeleton may be considered as a graph, with vertices and edges having their usual meanings. On the other hand, we can represent each face of the polyhedron by a vertex of the graph, and join two of these by an edge if and only if the faces they represent share a common edge.

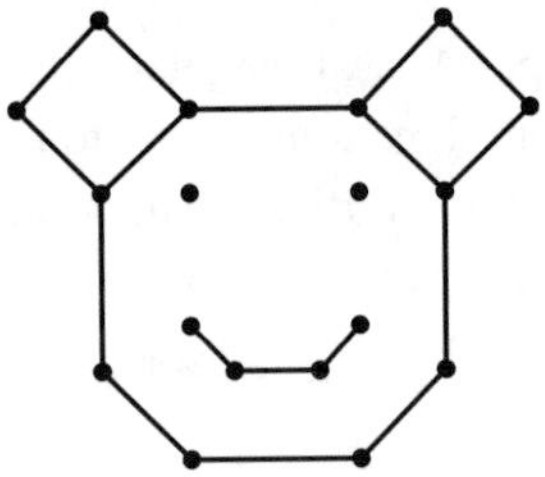

FIGURE 2.1.1

We often represent a graph in diagram form as a bunch of dots and lines such as the bear in Figure 2.1.1. Note that the diagram is only an aid to visualization, and can be drawn in many different ways as long as the same pairs of vertices are joined by edges. Whether the edges are straight and how long they are have no bearing on the structure of the graph.

A **path** is defined as an alternating sequence $v_0, e_1, v_1, e_2, v_2, \ldots, e_n, v_n$ of distinct vertices and edges such that for $1 \leq i \leq n$, the edge e_i joins the vertices v_{i-1} and v_i. The path is then said to *join* the vertices v_0 and v_n. A **cycle** is defined in the same way except that v_n is replaced by v_0 and e_n joins v_{n-1} and v_0. The ears and the face of the bear are cycles, and there are others. The **length** of a path or cycle is the number of edges on it. Note that the minimum length of a cycle is three.

A graph is said to be **connected** if every two vertices are joined by at least one path. Thus the bear in Figure 2.1.1 is not a connected graph. The eyes are isolated vertices, and in any case, the mouth is not connected to the head. Most of the graphs we encounter in this chapter will be connected.

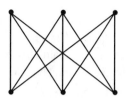

FIGURE **2.1.2**

The first two graphs in Figure 2.1.2 are **isomorphic**. This means that they represent the same graph. We can set up a one-to-one correspondence between their vertices so that two vertices in one are joined by an edge if and only if the corresponding two vertices in the other are joined by an edge.

Thus a graph can be drawn in many different ways. The second drawing may be preferable to the first, since edges do not cross but only meet at vertices. Note that the crossing in the first drawing does not constitute a fifth vertex. If a graph can be drawn without crossing edges, it is said to be **planar**. The third graph in Figure 2.1.2 is called the Utility Graph because of the following puzzle.

There are three feuding households in a small community which has its own supply of water, gas and electricity. Show how each family can run a pipe along the ground to each utility, so that no two pipes cross.

You may enjoy proving that the Utility Graph is not planar. How then can the puzzle above have a solution? Figure 2.1.3 is found in the answer section of a puzzle book.

This solution is sneaky and not realistic. Feuding families who do not even want their pipes to cross are hardly likely to allow the pipes of the others to run through their own homes. Similarly, in graph theory, we do not allow edges to run through vertices without terminating there.

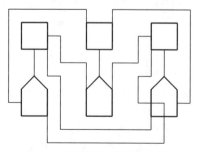

FIGURE **2.1.3**

The **degree** of a vertex is the number of edges having this vertex as an endpoint. The vertex is said to be **odd** or **even** as its degree is odd or even. A graph is said to be **regular** of degree k if k is the degree of each vertex. A regular graph of degree 1 consists of an even number of vertices joined in pairs by independent edges.

The **Graph Parity Theorem** states that the sum of degrees of all vertices in a graph is even. In fact, it is equal to twice the number of edges of the graph. This is because an edge contributes 1 to the degree of each of its endpoints. Since it has two endpoints, its total contribution is 2.

The **Cycle Decomposition Theorem** states that a regular graph of degree 2 is a union of disjoint cycles. To see this, we attempt to trace the whole graph, starting from any vertex. We trace either of its two edges which takes us to another vertex. It has only one edge along which we can continue the tracing. This process can only terminate at the vertex from which we started. This completes a cycle, and no vertex on it is joined to any vertices not on the cycle. If we have not traced the whole graph, we can start another cycle from a vertex not on the first one.

There is a special class of graphs called **bipartite graphs**. In such a graph, the vertices are partitioned into two sets, and there are no edges joining vertices within the same set. For example, the Utility Graph we met earlier is a bipartite graph.

Bipartite graphs are particularly useful as models. For example, the vertices in one set may represent readers of a journal who have submitted solutions to problems. The vertices in the other set may represent problems in the journal. An edge joins vertex u in the first set to vertex v in the second set if and only if u has submitted a solution to v.

The **Bipartite Cycle Theorem** states that every cycle in a bipartite graph has even length. This is easy to see since the vertices come alternately from the two sets. Since the cycle ends where it begins, there is an even number of vertices on it. It follows that the number of edges on the cycle is also even. Conversely, if every cycle of a graph has even length, then the graph must be bipartite.

2.1.3   Enumeration Techniques

A few of the problems ask for the counting of the elements in certain finite sets. We will denote the number of elements in a set S by $|S|$. We now present some of the basic counting techniques.

Let A and B be two sets. Their **Cartesian product** $A \times B$ is the collection of all ordered pairs (a, b) where a is an element in A and b is an element in B. Two ordered pairs are distinct if they differ in at least one component. The **Multiplication Principle** states that $|A \times B| = |A| \cdot |B|$. It holds because each of the $|A|$ elements of A can be paired with each of the $|B|$ elements of B.

For example, we wish to find pairs of positive integers (a, b) such that a is a multiple of 3 not greater than 15, and b is a multiple of 5 not greater than 15. Since the numbers involved are fairly small, we can list all the ordered pairs. They are, in ascending order of $a + b$: (3,5), (6,5), (3,10), (9,5), (6,10), (12,5), (3,15), (9,10), (15,5), (6,15), (12,10), (9,15), (15,10), (12,15) and (15,15). There are 15 ordered pairs.

This approach is called tackling a problem by brute force, or the brutal method. Apart from being time-consuming, there is always the nagging doubt whether we have omitted something or listed something more than once. The art of counting is knowing how to count without actually doing the counting.

We know that a is an element of the set $A = \{3, 6, 9, 12, 15\}$, and that b is an element of the set $B = \{5, 10, 15\}$. By the Multiplication Principle, the total number of ordered pairs (a, b) is $5 \times 3 = 15$. We do not even have to list the elements of A and B. Since every third number is a multiple of 3, $|A| = \frac{15}{3} = 5$. Since every fifth number is a multiple of 5, $|B| = \frac{15}{5} = 3$.

The **union** of two sets A and B is the set consisting of all elements in either A or B, and is denoted by $A \cup B$. In the example above, $A \cup B = \{3, 5, 6, 9, 10, 12, 15\}$, so that $|A \cup B| \leq |A| + |B|$ in general. In this example, 15 belongs to both both sets. The **Addition Principle** states that if A and B have no elements in common, then

$$|A \cup B| = |A| + |B|.$$

Note that both principles are indirect counting methods, in that they simply express the numbers of elements in certain sets in terms of the numbers of elements in other sets. At the end of the day, we still have to do some direct counting, such as the numbers of multiples of certain integers in certain sets.

We now consider a more difficult problem. *Consider the set of all triangles with integer sides. Find the number of those with perimeter 15.* Let the side lengths be a, b and c. We may name the sides so that $a \leq b \leq c$, and denote this triangle by (a, b, c). The problem specifies that $a + b + c = 15$. Of course, we must still have $a + b > c$ because of the Triangle Inequality.

The number 15 is not too large. If during a competition you cannot think of a way to tackle this problem other than the brutal method, you may as well go for it. If you persevere with care, you should succeed and be rewarded. After the contest, you should continue to search for a better method that offers a deeper understanding of the problem.

Sometimes, even in a competition, it is easier to work with the general case rather than a specific one. On the other hand you can build up a solution from simpler cases. Here, we shall replace the number 15 by an integer variable n, and solve the problem for some small values of n. We denote the number of triangles we seek by f_n, so that the original problem asks for the value of f_{15}.

Clearly,

$$f_0 = f_1 = f_2 = 0 \quad \text{and} \quad f_3 = 1,$$

the only triangle with perimeter 3 being (1,1,1). Then we have

$$f_4 = 0 \quad \text{and} \quad f_5 = f_6 = 1,$$

these triangles being (1,2,2) and (2,2,2). Note that the second one can be obtained from (1,1,1) by adding 1 to each side length. Since $a + b > c$ implies $(a+1) + (b+1) > c + 2 > c + 1$, this works in general.

However, this process is not always reversible. For example, if we subtract 1 from each side length of (2,3,4), we will end up with a degenerate triangle of side lengths 1, 2 and 3. Thus (2,3,4) is an *irreducible* triangle.

Note that since a, b and c are positive integers, $a + b > c$ implies $(a-1) + (b-1) \geq c - 1$. If (a,b,c) is an irreducible triangle, then $(a-1) + (b-1) = c - 1$ or equivalently $a + b = c + 1$. So it barely satisfies the Triangle Inequality.

It would appear that if we can find all the irreducible triangles, we should have an easy time with the original problem. So let us denote by g_n the number of irreducible triangles (a, b, c) with perimeter n. For these triangles, $a + b + c = n$ and $a + b = c + 1$.

At this point, we introduce the term *parity*. The parity of an integer is even if it is a multiple of 2, and odd if it is not a multiple of 2. From this definition, we deduce that the sum of two even integers is even, the sum of two odd integers is also even, while the sum of an odd and an even integer is odd. Moreover, an odd integer can never be equal to an even integer. We refer to these results collectively as the **Parity Principle**.

Since $2c$ is even and 1 is odd, the Parity Principle tells us that $n = a + b + c = 2c + 1$ is odd. Hence there are no irreducible triangles with even perimeter, and $g_n = 0$ when n is even.

Suppose n is odd. Then $a + b = c + 1 = \frac{n-1}{2} + 1 = \frac{n+1}{2}$. If $\frac{n+1}{2}$ is even, say $\frac{n+1}{2} = 2k$, then $n = 4k - 1$. Since $a \leq b$, a can have any of the values of $1, 2, \ldots, \frac{n+1}{4} = k$. If $\frac{n+1}{2}$ is odd, say $\frac{n+1}{2} = 2k + 1$, then $n = 4k + 1$ and a can have any of the values $1, 2, \ldots, \frac{n-1}{4} = k$. In summary, $g_{4k} = g_{4k+2} = 0$ and $g_{4k-1} = g_{4k+1} = k$.

We can now solve the original problem. There are two kinds of triangles counted in f_n. The number of irreducible ones is g_n, while the number of reducible ones is f_{n-3}. This is because every triangle with perimeter $n-3$ gives rise to a reducible triangle with perimeter n, while every reducible triangle with perimeter n can be reduced to a triangle with perimeter $n-3$. This is known as a **one-to-one correspondence**.

It follows that
$$f_n = g_n + f_{n-3}; \quad n > 3.$$

This is known as a **recurrence relation** for the sequence $\{f_n\}$. Since we know the values of g_n, the values of f_n for large n can be obtained from those for small n. We may rewrite the recurrence relation as follows:

$$f_n - f_{n-3} = g_n,$$
$$f_{n-3} - f_{n-6} = g_{n-3},$$
$$f_{n-6} - f_{n-9} = g_{n-6},$$
$$\cdots = \cdots.$$

When we add these, all but two terms on the left side cancel out. Such a sum is known as a *telescoping sum*. We have

$$f_{15} = f_{15} - f_0 = g_{15} + g_{12} + g_9 + g_6 + g_3 = 4 + 0 + 2 + 0 + 1 = 7.$$

2.1.4 Finite and Infinite Sets

Most combinatorics problems deal with finite sets, though **Problem 1934.3** involves an infinite set. The finiteness of a set gives rise to many of its basic properties. We present some of the most important results. Let us assume that all sets we deal with here are *non-empty*, unless explicitly stated otherwise.

The union of finitely many finite sets is also finite. In other words, if the union of finitely many sets is infinite, then at least one of those sets must also be infinite. We shall refer to this result as the **Finite Union Principle**. As a corollary, note that if we remove finitely many elements from an infinite set, then infinitely many elements are left behind. This is because the original set is the union of the set of those removed and the set of those left behind.

A **multi-set** is very much like a set except that it can contain identical elements. The number of times an element appears is called its *multiplicity*.

For instance, the multi-set $S = \{78, 92, 92, 64, 68, 84\}$ may consist of the scores of six students in a test.

An important property of finite multi-sets of real numbers is the **Extremal Value Principle**. It states that every finite multi-set of real numbers has a maximum and a minimum. Note that the maximum 92 of S has multiplicity two. In the discussion of the Power Means Inequality earlier, we have presupposed the existence of the maximum and the minimum. This assumption is now justified by the Extremal Value Principle.

Note that we could get by with postulating the existence of just a maximum or just a minimum. The existence of the other can be deduced simply by switching the signs of all the numbers and then applying the previous result. For example, the minimum of the multi-set S of test scores is 64 where -64 is the maximum of the multi-set $\{-78, -92, -92, -64, -68, -84\}$. However, we prefer to state this principle in the symmetric form.

We can give a proof of the existence of a maximum as follows. Pick any two of the numbers and throw away the smaller one, or either one if they are equal. Repeat this process. Since the set of numbers is *finite*, the process eventually terminates, and the number we still have is a maximum. Applying this to the elements of the multi-set S in the order listed, we reject in turn 78, 92, 64, 68 and 84, leaving 92 as the maximum.

We can solve a one-dimensional version of **Problem 1935.2** in the same way. By comparing the points in a finite set S two at a time, we can pick out the one that is furthest to the "left" and the one that is furthest to the "right." If S has a center of symmetry of S then it must be the point halfway between these two extremal points.

The **Mean Value Principle** states that in every finite multi-set of real numbers, there is at least one that is not less than the arithmetic mean of the set, and at least one not greater. This follows from the Extremal Value Principle and the Power Means Inequality. For instance, the arithmetic mean of the multi-set S of test scores is $79\frac{2}{3}$, which is less than the maximum 92 and greater than the minimum 64.

In some sense, the Finite Union Principle is an infinite version of the Mean Value Principle. Here we have infinitely many elements and finitely many sets. Hence the "average" number of elements per set is infinite, and at least one set must have infinitely many elements.

A most important corollary of the Mean Value Principle is the **Pigeonhole Principle**. Finitely many pigeons are put into finitely many holes. If there are more pigeons than holes, then at least one hole contains at least two pigeons. If there are more holes than pigeons, then there is at least one empty hole.

If there are more pigeons than holes, then the average number of pigeons per hole is greater than one. It is understood, especially by the pigeons, that they are not to be carved up. By the Mean Value Principle, there is at least one hole containing no less than the average number of pigeons. This means at least two pigeons are in it. Similarly, if there are more holes than pigeons, then the average number of pigeons is less than one, which means that the number of pigeons in some hole must be zero since we are dealing with non-negative integers.

The following problem bears some superficial resemblance to **Problem 1943.1**. *Prove that in any group of people, there must be two who know the same number of the others in the group. Assume that "knowing" is a symmetric relation.*

The key phrases are "must be two" and "the same." These are strong hints that the Pigeonhole Principle may be involved. Clearly, the people should be the pigeons, and two go into the same pigeonhole if they know the same number of the others.

Let the total number of people be n. Then each may know 0, 1, 2, ..., $n - 1$ others. It appears that we have n pigeons and n pigeonholes, which is not a desirable scenario. However, if somebody knows no others, then nobody can know all the others. Hence we may eliminate either the pigeonhole for people who know 0 others or the pigeonhole for people who know $n - 1$ others. Now we have n pigeons and $n - 1$ pigeonholes, and the desired conclusion follows from the Pigeonhole Principle.

The Extremal Value Principle and its corollaries do not hold if we remove the assumption that the multi-set of numbers we are dealing with is finite. For example, the set of all integers does not have either a maximum or a minimum. However, this set has a special property not shared by the set of real numbers.

Given an arbitrary integer, say 54, we can say that 53 is the next integer below 54, and 55 is the next integer above 54. If we regard 54 as a real number, then we cannot answer the questions of which real number is the next above 54, and which real number is the next below 54. If someone claims that some $x > 54$ is the next real number above 54, we can point out that $\frac{x+54}{2}$ would be a better choice.

Let us state this special property of the set of integers properly. We need two definitions. A set S of real numbers is said to be **bounded above** if there exists a real number M such that $x \leq M$ for every $x \in S$. It is said to be **bounded below** if there exists a real number m such that $x \geq m$ for every $x \in S$. Note that the Extremal Value Principle can fail for an infinite set of real numbers even if the set is bounded. For example, the set of reciprocals of all positive integers does not have a minimum.

The **Well-Ordering Principle** states that a *non-empty* set of integers that is bounded above has a maximum, and a *non-empty* set of integers that is bounded below has a minimum. Since the set of integers greater than 54 is non-empty and bounded below by 54, it has a minimum. This minimum is the next integer after 54. Of course, that number is 55.

2.2   Solutions

Problem 1940.1.

In a set of objects, each has one of two colors and one of two shapes. There is at least one object of each color and at least one object of each shape. Prove that there exist two objects in the set that are different both in color and in shape.

First Solution: Let the colors be red and green. If we have red objects of both shapes, one of them will form a mismatching pair with some green object. If all the red objects are of the same shape, then there must be a green object of the other shape. We will also have a mismatching pair.

Second Solution: Let the colors be red and green, and the shapes be square and circular. We may assume that no two objects are identical in color as well as in shape. If there is only one, say a red square object, then we have no green ones and no circular ones. If there are exactly two objects and we have objects of both colors and both shapes, these two objects must differ in color and in shape. Henceforth, we may assume that there are at least three objects.

Construct a bipartite graph as shown in Figure 2.2.1. The vertices R and G on one side represent the colors, while the vertices S and C on the other side represent the shapes. Each edge represents one kind of object. We shall apply the Pigeonhole Principle as follows. The objects are the pigeons. A pigeon goes into the first pigeonhole if it is represented by the edge RS or the edge GC, and into the second if represented by RC or GS. Since there are exactly two pigeonholes and at least three pigeons, two of the pigeons will be in the same pigeonhole. These two objects will differ in color and in shape.

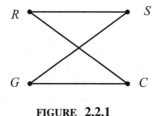

FIGURE 2.2.1

Problem 1943.1.

Prove that in any group of people, the number of those who know an odd number of the others in the group is even. Assume that "knowing" is a symmetric relation.

First Solution: We will host a party for this group and ask them to shake hands with those they know. The handshakes are to be performed one at a time before us, so that we can keep track of how many hands each has shaken so far. Each guest is considered "odd" or "even" at the time according to the parity of this number. In a handshake between two odd guests, both will become even. In a handshake between two even guests, both will become odd. In a handshake between an odd guest and an even guest, they exchange status. Hence the number of odd guests changes by -2, 2 or 0, all of which are even numbers. Before any handshakes took place, there were no odd guests, and 0 is an even number. It follows that after any number of handshakes, and in particular after all handshakes, the number of odd guests is still even. Since we arranged that only people who know each other shake hands, the number of those who know an odd number of the others in the group is even.

Second Solution: We construct a graph as follows. Each vertex represents a person in the group and two vertices are joined by an edge if the people they represent know each other. What we have to prove is that the number of odd vertices is even. Suppose this is not the case. We shall derive a contradiction via the Parity Principle. Since we assume that there is an odd number of odd vertices, the sum of their degrees must be odd. On the other hand, the sum of the degrees of the even vertices is always even. Hence the sum of the degrees of all the vertices is odd. However, this contradicts the Graph Parity Theorem. It follows that the number of people who know an odd number of the others in the group is even.

Problem 1933.2.

Sixteen squares of an 8×8 chessboard are chosen so that there are exactly two in each row and two in each column. Prove that eight white pawns and eight black pawns can be placed on these sixteen squares so that there is one white pawn and one black pawn in each row and in each column.

First Solution: Put a white pawn on any of the given squares. There is exactly one other given square on the same row, and we put a black pawn on it. Now there is exactly one other given square on the same column as this square, and we put a white pawn on it. We continue this process until our trail leads us back to the square chosen initially. It cannot lead to another square which we have already visited because each square has exactly two access routes. Since the moves are alternately horizontal and vertical, we have made an even number of moves and placed an even number of pawns that alternate in color. If all the given squares have pawns on them, the task is completed. If not, we repeat the process starting with any of the given squares still without a pawn. The new trail cannot intersect the old one at a vertex, again because each square has exactly two access routes. Using additional trails if necessary, we can eventually put a pawn on each of the given squares according to the stipulation. Since there are finitely many such squares, the process must terminate.

Second Solution: We construct a bipartite graph as follows. On the left, we have eight vertices representing the rows of the chessboard. On the right, we have eight vertices representing the columns of the chessboard. An edge joining a vertex on each side represents a given square at the intersection of the row and column represented by the vertices. Thus there are exactly 16 edges, and the graph is regular of degree 2. By the Cycle Decomposition Theorem, it is a disjoint union of cycles. By the Bipartite Cycle Theorem, all of them have even lengths. Thus we can color the edges of each cycle alternately black and white. Then we place a pawn of the corresponding color on the given square represented by each edge. Since two given squares in the same row or the same column are represented by adjacent edges, the pawns on them must be of opposite colors.

Problem 1930.2.

A straight line is drawn across an 8×8 chessboard. It is said to pierce a square if it passes through an interior point of the square. At most how many of the 64 squares can this line pierce?

Solution:

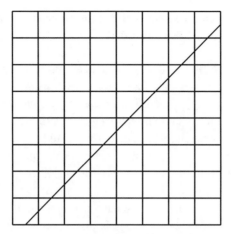

FIGURE 2.2.2

Internally, the chessboard has seven horizontal and seven vertical grid lines. If we move along an arbitrary line from one square to another, we must cross one of the 14 grid lines. Since two lines intersect at most once, there are at most 14 crossings. It follows that a line can pierce no more than 15 squares. Figure 2.2.2 shows that there is a line that pierces 15 squares.

Remark: On an $m \times n$ chessboard, the answer is $(m-1)+(n-1)+1 = m+n-1$.

Problem 1930.1.

How many five-digit multiples of 3 end with the digit 6?

First Solution: The number of five digit numbers is $99999 - 9999 = 90000$. Every third of them is divisible by 3. Hence the number of five-digit multiples of 3 is $90000 \div 3 = 30000$. The multiples of 3 end periodically in 3, 6, 9, 2, 5, 8, 1, 4, 7 and 0. Hence the number of five-digit multiples of 3 that end with the digit 6 is $30000 \div 10 = 3000$.

Second Solution: A five-digit number ending in 6 is divisible by 3 if and only if the four-digit number obtained by removing this 6 is divisible by 3. There are $9999 - 999 = 9000$ four-digit numbers, and every third is divisible by 3. Hence the number of five-digit multiples of 3 that end with the digit 6 is $9000 \div 3 = 3000$.

Problem 1929.1.

In how many ways can the sum of 100 fillér be made up with coins of denominations 1, 2, 10, 20 and 50 fillér?

First Solution: We first solve two preliminary problems. Let f_n denote the number of ways of making up a sum of n fillér using only 1-fillér and 2-fillér coins. We have $f_{2k} = k + 1$ since we can use anywhere from 0 to k 2-fillér coins, and make up the remaining sum with 1-fillér coins. Also, $f_{2k+1} = k + 1$ since we have no alternative to adding a 1-fillér coin to any collection with total value $2k$ fillér. Let g_n denote the number of ways of making up a sum of n fillér using 1-fillér, 2-fillér and hypothetical 5-fillér coins. We define $g_n = 0$ if $n < 0$. For $n \geq 0$, we claim that $g_n = f_n + g_{n-5}$. This is because we have two options. If we do not use any 5-fillér coins, the number of ways is just f_n. If we use at least one 5-fillér coin, then the remaining sum of $n - 5$ fillér can be made up using 1-fillér, 2-fillér and more 5-fillér coins. The Addition Principle justifies our claim. We now make up a sum of 100 fillér using 1-fillér, 2-fillér, 10-fillér, 20-fillér and 50-fillér coins. Note that the 1-fillér and 2-fillér coins together must make up a sum of $10i$ fillér for some $i, 0 \leq i \leq 10$. There are f_{10i} ways of doing this. The remaining sum $100 - 10i = 10(10 - i)$ fillér is to be made up with 10-fillér, 20-fillér and 50-fillér coins. There are g_{10-i} ways of doing that. This is equivalent to making up a sum of $10 - i$ fillér using 1-fillér, 2-fillér and 5-fillér coins. By the Multiplication Principle, the number of ways for a particular i is $f_{10i}g_{10-i}$. Since two ways with different values of i are distinct, the Addition Principle tells us that the total number of ways is the sum of all the numbers in the last row of the following table, which is 784.

i	0	1	2	3	4	5	6	7	8	9	10
f_i	1	1	2	2	3	3	4	4	5	5	6
g_i	1	1	2	2	3	4	5	6	7	8	10
f_{10i}	1	6	11	16	21	26	31	36	41	46	51
g_{10-i}	10	8	7	6	5	4	3	2	2	1	1
$f_{10i}g_{10-i}$	10	48	77	96	105	104	93	72	82	46	51

Second Solution: We count the number of ways directly and systematically. Start with the coin in the highest denomination and use as many of it as we can. If the total is not exactly 100 fillér, we go to the coin in the next highest denomination, and so on. We do not worry about 1-fillér coins since the number used is determined by the combination of other coins used. Then we reduce by 1 at a time the number of coins used in reverse denominational order. The process is summarized in the following table. The final answer, 784, is the sum of all the numbers in the last column.

Number of coins used				Number
50-fillér	20-fillér	10-fillér	2-fillér	of ways
2	0	0	0	1
1	2	1	0	1
		0	0–5	6
	1	3	0	1
		⋮	⋮	⋮
		0	0–15	16
	0	5	0	1
		⋮	⋮	⋮
		0	0–25	26
0	5	0	0	1
	4	2	0	1
		1	0–5	6
		0	0–10	11
	3	4	0	1
		⋮	⋮	⋮
		0	0–20	21
	2	6	0	1
		⋮	⋮	⋮
		0	0–30	31
	1	8	0	1
		⋮	⋮	⋮
		0	0–40	41
	0	10	0	1
		⋮	⋮	⋮
		0	0–50	51

Problem 1935.3.

A real number is assigned to each vertex of a triangular prism so that the number on any vertex is the arithmetic mean of the numbers on the three adjacent vertices. Prove that all six numbers are equal.

First Solution:

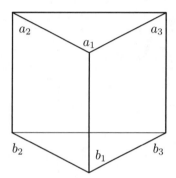

FIGURE 2.2.3

Let the numbers be as shown in Figure 2.2.3. Then we have

$$3a_1 = a_2 + a_3 + b_1, \tag{1}$$

$$3a_2 = a_3 + a_1 + b_2, \tag{2}$$

$$3a_3 = a_1 + a_2 + b_3, \tag{3}$$

$$3b_1 = b_2 + b_3 + a_1, \tag{4}$$

$$3b_2 = b_3 + b_1 + a_2. \tag{5}$$

Now $(1) + (2) + (3)$, $(1) + (4)$ and $(2) + (5)$ yield

$$a_1 + a_2 + a_3 = b_1 + b_2 + b_3, \tag{6}$$

$$2(a_1 + b_1) = (a_2 + b_2) + (a_3 + b_3), \tag{7}$$

$$2(a_2 + b_2) = (a_3 + b_3) + (a_1 + b_1). \tag{8}$$

Next, $(7) - (8)$ yields

$$a_1 + b_1 = a_2 + b_2. \tag{9}$$

Finally, $(1) - (2) + (9)$ yields $a_1 = a_2$. By symmetry, $a_1 = a_2 = a_3$ and $b_1 = b_2 = b_3$. It follows from (6) that $a_1 = a_2 = a_3 = b_1 = b_2 = b_3$.

Remark: Since the problem is unaffected if we add or subtract the same number from the numbers on all the vertices, we may simply take $a_1 = 0$.

Second Solution: We prove more generally that the result holds for any polyhedron whose skeleton is a connected graph. By the Extremal Value Principle, the number m on one of the vertices is the maximum among these numbers. Since it is the arithmetic mean of its neighbors, those numbers are all equal to m by the equality clause of the Power Means Inequality. Since the graph is connected, this value m will spread to all vertices.

Problem 1935.2.

Prove that a finite point set cannot have more than one center of symmetry.

First Solution: Let O be a center of symmetry of the finite point set S. By the Extremal Value Principle, there exists a point A such that $OA \geq OP$ for any point P in S. Note that the point B symmetric to A with respect to O is also in S.

Suppose O' is another center of symmetry of S. Then the point B' symmetric to A with respect to O' is also in S, so that $OB' \leq OA$. We may assume that $O'A \geq O'B$. Suppose O' lies on the line AB. Then O must be between A and O', so that O' is between O and B'. Hence $OB' > O'B' = O'A > OA$, which is a contradiction. Suppose O' is not on the line AB, as illustrated in Figure 2.2.4. By the Triangle Inequality, $2O'A \geq O'A + O'B > AB = 2OA$, so that $OB' \geq AB' - OA = 2O'A - OA > OA$. This is also a contradiction.

Remark: For other solutions to this problem, see pages 68 and 74 in Chapter 5, as well as page 114 in Chapter 6.

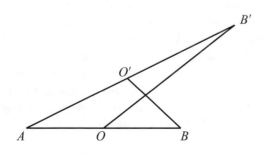

FIGURE 2.2.4

Problem 1934.3.

We are given an infinite set of rectangles in the plane, each with vertices of the form $(0,0)$, $(0,m)$, $(n,0)$ and (n,m), where m and n are positive integers. Prove that there exist two rectangles in the set such that one contains the other.

First Solution: We assume that the result is not true. The widths of the rectangles are positive integers. By the Well-Ordering Principle, there is a rectangle whose width n_0 is minimum. Then its height m_0 must be maximum as otherwise it will be contained in some other rectangle. We can continue to choose a rectangle that is the next narrowest as well as the next tallest. However, since the height is a decreasing positive integer, the process must stop at some point. This contradicts the hypothesis that the set of rectangles is infinite. It follows that the desired result holds.

Second Solution: The widths of the rectangles are positive integers. By the Well-Ordering Principle, there is a rectangle R whose width n is minimum. Let m be the height of this rectangle and consider any m of the other rectangles. If any of them has height greater than or equal to m, then it contains R. If not, then their heights range from 1 to $m - 1$. By the Pigeonhole Principle, two of them will have the same height, and one will contain the other.

Third Solution: We claim that there is an infinite sequence of these rectangles such that each is contained in the next. The desired conclusion will follow from this stronger result. To justify the claim, we consider two cases. First, suppose that every rectangle is contained in some other rectangle. Then we can build up this sequence by starting with any rectangle, and there is always a choice for the next one. Now suppose that there is a rectangle R not contained in any others. Then every other rectangle is either shorter or narrower. There are infinitely many of them. By the Finite Union Principle, there are either infinitely many rectangles that are shorter or infinitely many that are narrower. By symmetry, we may assume that infinitely many are shorter. Now there are finitely many possible heights of these rectangles. By the Finite Union Principle again, infinitely many of them have the same height. If we arrange them in increasing order of width, we have the desired infinite sequence.

Fourth Solution: If two rectangles have the same height or the same width, then one will contain the other. Hence we may assume that the heights are distinct and so are the widths. Take any $m \times n$ rectangle R. There are at most $m-1$ other rectangles shorter than R, and at most $n-1$ rectangles narrower than R. Hence there are finitely many such rectangles. By the Finite Union Principle, infinitely many rectangles remain, and each of them contains R.

3
Number Theory Problems

Problem 1938.1.

Prove that an integer n can be expressed as the sum of two squares if and only if $2n$ can be expressed as the sum of two squares.

Problem 1942.2.

Let a, b, c and d be integers such that for all integers m and n, there exist integers x and y such that $ax + by = m$ and $cx + dy = n$. Prove that $ad - bc = \pm 1$.

Problem 1931.1.

Let p be a prime greater than 2. Prove that $\frac{2}{p}$ can be expressed in exactly one way in the form $\frac{1}{x} + \frac{1}{y}$ where x and y are positive integers with $x > y$.

Problem 1940.2.

Let m and n be distinct positive integers. Prove that $2^{2^m} + 1$ and $2^{2^n} + 1$ have no common divisor greater than 1.

Problem 1931.2.

Let $a_1^2 + a_2^2 + a_3^2 + a_4^2 + a_5^2 = b^2$, where a_1, a_2, a_3, a_4, a_5 and b are integers. Prove that not all of these numbers can be odd.

Problem 1939.2.

Determine the highest power of 2 that divides $2^n!$.

Problem 1934.1.

Let n be a given positive integer and

$$A = \frac{1 \cdot 3 \cdot 5 \cdots (2n-1)}{2 \cdot 4 \cdot 6 \cdots 2n}.$$

Prove that at least one term of the sequence $A, 2A, 4A, 8A, \ldots, 2^k A, \ldots$ is an integer.

Problem 1932.1.

Let a, b and n be positive integers such that b is divisible by a^n. Prove that $(a+1)^b - 1$ is divisible by a^{n+1}.

Problem 1936.3.

Let a be any positive integer. Prove that there exists a unique pair of positive integers x and y such that $x + \frac{1}{2}(x+y-1)(x+y-2) = a$.

3.1 Discussion

3.1.1 Mathematical Induction

In Chapter 2, we came across the **Well-Ordering Principle**. While its statement may seem simple, it is a very powerful tool. It is particularly useful in proving an assertion that a statement $P(n)$, which depends on the variable n, is true for all positive integers n. We can possibly prove a few cases for small values of n, but we need some general strategy to prove the entire sequence of statements.

Suppose $P(n)$ is not true for all n. Then the set S of positive integers n for which $P(n)$ is false is non-empty. Since S is bounded below by 0, it has a minimum m. Now $P(m)$ is false while $P(n)$ is true for all positive integers $n < m$. What we need to do is to derive from the falsehood of $P(m)$ that $P(n)$ is also false for some positive integer $n < m$. Then we will have a contradiction. The most likely candidate for the counterexample is $n = m - 1$, but there are other possible ones such as $n = \frac{m}{2}$ if m is even.

Let $P(n)$ be the statement $1 + 2 + 2^2 + \cdots + 2^{n-1} = 2^n - 1$. When $n = 1$, the first term on the left is 1 and the last is $2^{1-1} = 1$. Hence the sum reduces to a single term 1. The expression on the right is equal to $2^1 - 1 = 1$. Hence $P(1)$ is true as asserted.

Suppose $P(n)$ is not always true. Then there exists a smallest positive integer m for which $P(m)$ is false, that is,

$$1 + 2 + 2^2 + \cdots + 2^{m-1} \neq 2^m - 1.$$

Subtracting 2^{m-1} from each side, we have

$$1 + 2 + 2^2 + \cdots + 2^{m-2} \neq 2^m - 1 - 2^{m-1} = 2^{m-1} - 1.$$

However, this means that $P(m-1)$ is also false. Since $m > 1$, $m-1$ is a positive integer. Thus the claim that m is the smallest positive integer for which $P(m)$ is false is invalid. It follows that the statement $P(n)$ must be true for all n.

We now derive from the Well-Ordering Principle an important result known as the **Principle of Mathematical Induction**. It states that if S is a set of positive integers such that $1 \in S$, and $n+1 \in S$ whenever $n \in S$, then S is the set of all positive integers.

To prove this principle, suppose S is not the set of all positive integers. Then the set T of positive integers not in S is non-empty. Moreover, since $1 \in S$, T is bounded below by 1. By the Well-Ordering Principle, T has a smallest element $m > 1$. Now $m - 1$ is a positive integer smaller than m. Hence $m - 1 \in S$, and it follows that $(m-1) + 1 = m \in S$ also. This is a contradiction.

We can also derive the Well-Ordering Principle from the Principle of Mathematical Induction, so that the two results are equivalent. It is sufficient to prove the special case which states that every non-empty set of positive integers has a minimum.

Let E be a set of positive integers without a minimum, and let S be the set of positive integers n such that all positive integers from 1 to n are not in S. Clearly, 1 is not in E, as otherwise it would have been its minimum. Hence 1 is in S. Suppose n is in S for some $n \geq 1$. Then all of 1, 2, ..., n are not in E. If $n + 1$ is in E, it would have been its minimum. Hence all of 1, 2, ..., $n + 1$ are not in E, so that $n + 1$ is in S. By the Principle of Mathematical Induction, S is the set of all positive integers. This means that every positive integer n in not in E. In other words, the set E of positive integers without a minimum can only be the empty set.

In using the Principle of Mathematical Induction to solve problems, we first prove that $P(1)$ is true, which is usually not too difficult. This initial case serves as the **basis** for induction. The key step is to prove that if $P(n)$ is true for *some* positive integer n, then so is $P(n + 1)$. Now let S be the set of all positive integers for which $P(n)$ is true. By what we have proved, $1 \in S$ and $n + 1 \in S$ whenever $n \in S$. By the Principle of Mathematical Induction, S is the set of all positive integers.

The second step is the essence of this method of mathematical induction. The assumption that $P(n)$ is true for some n is called the **induction hypothesis**. Note that we are not assuming what we have to prove, namely, $P(n)$ is true for *all* positive integers n. After establishing the basis, we are certainly justified to make the induction hypothesis.

We now solve the earlier problem about the sum of the powers of 2. The basis that $P(1)$ is true has already been established. Suppose that $P(n)$ is true for some positive integer n, that is,

$$1 + 2 + 2^2 + \cdots + 2^{n-1} = 2^n - 1.$$

Adding 2^n to each side, we have

$$1 + 2 + 2^2 + \cdots + 2^{(n+1)-1} = 2^{n+1} - 1.$$

Hence $P(n + 1)$ is also true. This completes the inductive step and the induction argument.

It may be noted that the two approaches are not substantially different. It boils down to a choice in presentation. However, it is possible to attack this particular problem in a completely different way, which is incidentally direct. Let

$$S = 1 + 2 + 2^2 + \cdots + 2^{n-1}.$$

Then

$$2S = 2 + 2^2 + 2^3 + \cdots + 2^n.$$

Subtracting the first equation from the second, we have $S = 2^n - 1$.

The derivation of the Well-Ordering Principle from the Principle of Mathematical Induction suggests a variation of the method of induction. The basis is that $P(1)$ is true as before. The induction hypothesis is modified so that if $P(1), P(2), \ldots, P(n)$ are true for some positive integer n, then so is $P(n+1)$. Another variation may be found in Chapter 4 in the proof of inequalities.

3.1.2 Divisibility

A positive integer b is said to be **divisible** by a positive integer a if there exists a positive integer q such that $b = aq$. We also say that a divides b, b is a multiple of a, and a is a divisor of b. Symbolically, we write $a|b$.

The binary relation of divisibility on the set of positive integers $\mathcal{N}$ has the following properties:

1. **Reflexivity:** $a \in \mathcal{N} \implies a|a$.
2. **Antisymmetry:** $a|b, b|a \implies a = b$.
3. **Transitivity:** $a|b, b|c \implies a|c$.

A binary relation with these three properties is called a **partial order** relation. For the remainder of this Section, all variables are assumed to be positive integers unless otherwise stated.

Many simple results on divisibility can be derived from the definition and the preceding properties. Let b be divisible by a. Then bk is also divisible by a for any k. By definition, there exists q such that $b = aq$. Since multiplication is associative, $bk = (aq)k = a(qk)$, so that bk is indeed divisible by a. Alternatively, note that $b|bk$. From $a|b$, we have $a|bk$ by transitivity.

The **Distributivity Theorem** states that if $a + b = c$ and d divides two of a, b and c, then it also divides the third. Suppose $d|a$ and $d|b$. Then $a = dq$ and $b = dp$ for some q and p. Now $c = a+b = dq+dp = d(q+p)$ since multiplication is distributive over addition. It follows that $d|c$. The other two cases can be proved in an analogous manner.

The **greatest common divisor** of two positive integers a and b, denoted by $\gcd(a,b)$, is the positive integer d such that $d|a$, $d|b$ and $d \geq c$ for any positive integer c that divides both a and b. This always exists, since the number 1 divides all positive integers and is always ready to serve as the greatest common divisor if no larger number can be found that divides both. If $\gcd(a,b) = 1$ then a and b are said to be **relatively prime**.

The **Linearity Theorem** states that the greatest common divisor d of a and b is expressible as a **linear combination** of these two numbers. In other words, there exist integers x and y, not necessarily positive, such that $d = ax + by$. An important consequence of this is that if $c|a$ and $c|b$, then $c|ax$ and $c|by$ by our earlier result, and $c|d$ by the Distributivity Theorem. Note that $c|d$ is a much stronger statement than $c \leq d$ in the definition of the greatest common divisor.

Consider the set S of all *positive* integers expressible as linear combinations $ax + by$ of a and b, where x and y range over all integers. S is non-empty since $a = a \cdot 1 + b \cdot 0$ is in S. By the Well-Ordering Principle, S has a minimum $d = ax + by$ for some x and y. As shown in the preceding paragraph, $c|d$ whenever $c|a$ and $c|b$. In order to conclude that d is the greatest common divisor of a and b, it remains to be shown that $d|a$. We will then have $d|b$ by symmetry.

Suppose that d does not divide a. We know from arithmetic that by performing the so-called "long division," we will get a quotient q and a remainder r which satisfies $0 < r < d$, such that $a = dq + r$. This process is known as the **Division Algorithm**. Now

$$r = a - dq = a - (ax + by)q = a(1 - xq) + b(-yq).$$

Hence r belongs to S, but it is smaller than d. This is a contradiction.

Another important consequence of the Linearity Theorem is the **Relatively Prime Divisibility Theorem**, which states that if $a|bc$ and a is relatively prime to b, then $a|c$. The first condition means that $bc = aq$ for some q. The second condition means that 1 is the greatest common divisor of a and b. By the Linearity Theorem, $1 = ax + by$ for some integers x and y, not necessarily positive. Then $c = c(ax + by) = a(cx + qy)$, so that $a|c$.

A **prime** is a positive integer that has exactly two positive divisors, namely 1 and itself. A positive integer with at least three positive divisors is said to be **composite**. Note that 1 is neither a prime nor a composite number.

A prime is relatively prime to any number that is not a multiple of it. It follows from the Relatively Prime Divisibility Theorem that if a prime p divides the product ab, then $p|a$ or $p|b$. This is known as the **Prime Divisibility Theorem**.

Primes are important building blocks of the positive integers. If a positive integer greater than 1 is not a prime, it can be expressed as a product of two factors each greater than 1. If either of the factors is not a prime, it can be decomposed further. Eventually, we have an expression of the original number as a product of primes.

The **Fundamental Theorem of Arithmetic** states that the decomposition of any positive integer greater than 1 into primes is unique, if the prime factors are to be arranged in non-descending order.

3.1.3   Congruence

The concept of parity makes a distinction between the odd integers and the even integers. These two classes are simply numbers that, when divided by 2, leave remainders of 1 and 0 respectively. We can generalize this idea to division by numbers greater than 2.

Let m be a positive integer. Two integers a and b, not necessarily positive, are said to be **congruent** modulo m if they leave the same remainder r when divided by m, where $0 \le r < m$. Symbolically, we write $a \equiv b \pmod{m}$, and $a \not\equiv b \pmod{m}$ if the relation is not true.

Congruences abound in everyday life since many things come in cycles. For example, the hour of the day is based on a congruence modulo 12 or 24, and the day of the week is based on congruence modulo 7. Many secret codes using the English alphabet are based on congruence modulo 26.

Congruence on the set of integers $\mathcal{Z}$ is a binary relation with the following properties:

1. **Reflexivity:** $a \in \mathcal{Z} \Longrightarrow a \equiv a \pmod{m}$.
2. **Symmetry:** $a \equiv b \pmod{m} \Longrightarrow b \equiv a \pmod{m}$.
3. **Transitivity:** $a \equiv b \pmod{m}$, $b \equiv c \pmod{m}$
   $\Longrightarrow a \equiv c \pmod{m}$.
4. **Additivity:** $a \equiv b \pmod{m}$, $c \equiv d \pmod{m}$
   $\Longrightarrow a + c \equiv b + d \pmod{m}$.
5. **Multiplicativity:** $a \equiv b \pmod{m}$, $c \equiv d \pmod{m}$
   $\Longrightarrow ac \equiv bd \pmod{m}$.

The first three properties follow directly from the definition. A binary relation with these three properties is called an **equivalence** relation.

A condition equivalent to $a \equiv b \pmod{m}$ is $m|(a-b)$. To prove this, suppose $a \equiv b \pmod{m}$, so that $a = mq + r$ where $0 \leq r < m$. It follows that $b = mp + r$ for the same r. Hence $a - b = m(q - p)$ or $m|(a-b)$. Conversely, suppose $m|(a-b)$, so that $a - b = mk$ for some k. Now $a = mq + r$ where $0 \leq r < m$. Hence $b = a - mk = m(q - k) + r$, so that $a \equiv b \pmod{m}$.

Using this, we can easily prove the Additivity and Multiplicativity of congruence. We shall present only the latter here. We have $a - b = mk$ and $c - d = mh$ for some integers k and h. It follows that $ac - bd = (ac - bc) + (bc - bd) = m(kc + hb)$, from which we can conclude that $ac \equiv bd \pmod{m}$.

Suppose that R is an equivalence relation on a set S, and a is an element of S. Denote by $C(a)$ the subset of S consisting of all elements equivalent to a. Suppose that b belongs to $C(a)$. We claim that $C(b) = C(a)$. To prove this, let x be any element in $C(a)$. Then we have xRa. Since b also belongs to $C(a)$, we have aRb. By Transitivity, xRb and x belongs to $C(b)$. We can prove in the same way that any x in $C(b)$ also belongs to $C(a)$, justifying the claim. It follows that an equivalence relation partitions the elements of the set into *equivalence classes*.

Congruence modulo m partitions the integers into m congruence classes, according to their remainders when divided by m. We have seen that congruence modulo 2 partitions the integers into even and odd numbers. An important application of congruence is to determine to which congruence classes the squares can belong. In congruence modulo 3, the squares are all in the classes 0 and 1. The situation is the same modulo 4. The squares modulo 10 are all in the classes 0, 1, 4, 5, 6 and 9.

3.1.4 More Combinatorics

The notation ! in the statement of **Problem 1939.2** signifies the **factorial function**. Let n be an arbitrary positive integer. We will define $n! = n(n-1)\cdots 3 \cdot 2 \cdot 1$. We also define $0! = 1$. An important class of enumeration problems in combinatorics deals with **permutations**. However, they did not arise in connection with Chapter 2.

A permutation of a set is an arrangement of its elements in some order. To form a permutation of n objects, we can choose any of the n elements as the first, any of the remaining $n - 1$ as the second, and so on. An application of the Multiplicative Principle shows that the total number of permutations is $n!$.

A **combination** of a set is a selection of a subset of its elements. If there are n objects and we wish to select k of them, the total number of combinations is denoted by $\binom{n}{k}$, and is pronounced as "n choose k."

If we choose the k objects one at a time, the Multiplication Principle shows that the number of ways in which we can accomplish the task is $n(n-1)(n-2)\cdots(n-k+1)$. However, each of the chosen subsets appears in all of its permuted forms, of which there are $k!$. Hence

$$\binom{n}{k} = \frac{n(n-1)(n-2)\cdots(n-k+1)}{k!} = \frac{n!}{k!(n-k)!}.$$

The numbers $\binom{n}{k}$ are also called the **binomial coefficients**. A binomial is a polynomial with two terms, and the simplest binomial is $1 + x$. The **Binomial Theorem** states that

$$(1+x)^n = \binom{n}{0} + \binom{n}{1}x + \cdots + \binom{n}{n}x^n.$$

Let us consider the product $(1 + x_1)(1 + x_2)\cdots(1 + x_n)$. When we expand it, we take from the ith factor either the term 1 or the term x_i. If we do not take any x_i at all, we have only $\binom{n}{0} = 1$ choice, so that the only constant term is 1. If we take only one x_i, we have $\binom{n}{1} = n$ choices, so that there are n linear terms, namely, $x_1 + x_2 + \cdots + x_n$. Eventually, when we take all x_i, we have only $\binom{n}{n} = 1$ choice, so that the only nth-order term is $x_1 x_2 \cdots x_n$. We can now obtain the Binomial Theorem by setting $x_1 = x_2 = \cdots = x_n = x$.

The binomial coefficients satisfy an important recurrence relation with two indices. **Pascal's Formula** states that for $1 \leq k \leq n - 1$,

$$\binom{n}{k} = \binom{n-1}{k-1} + \binom{n-1}{k}.$$

This can be proved by using the expansion of the binomial coefficients into factorials. Indeed,

$$\binom{n-1}{k-1} + \binom{n-1}{k} = \frac{(n-1)!}{(k-1)!(n-k)!} + \frac{(n-1)!}{k!(n-k-1)!}$$

$$= \frac{(n-1)!}{k!(n-k)!}\left[k + (n-k)\right]$$

$$= \frac{n!}{k!(n-k)!} = \binom{n}{k}.$$

We offer a combinatorial proof of Pascal's Formula. Note that $\binom{n}{k}$ is the number of ways of choosing k elements from a set of n elements. We divide these choices into two classes according to whether a specific element is chosen. If it is, then there are $\binom{n-1}{k-1}$ ways of choosing the other $k - 1$ elements from the remaining $n - 1$. If it is not, then there are $\binom{n-1}{k}$ ways of choosing the k elements again from the remaining $n - 1$. Since these two classes are exhaustive and mutually exclusive, we have the desired result.

The binomial coefficients may be arranged in a famous configuration known in the west as **Pascal's Triangle**, though it had appeared in much earlier work in the Orient, and was known to Omar Khayyam. The first four rows are shown in Figure 3.1.1.

$$\binom{0}{0}$$
$$\binom{1}{0} \qquad \binom{1}{1}$$
$$\binom{2}{0} \qquad \binom{2}{1} \qquad \binom{2}{2}$$
$$\binom{3}{0} \qquad \binom{3}{1} \qquad \binom{3}{2} \qquad \binom{3}{3}$$

FIGURE 3.1.1

Since $\binom{n}{0} = 1 = \binom{n}{n}$, the numbers on the two slant sides of Pascal's Triangle are each 1. By Pascal's Formula, each interior number is the sum of the two in the row above that are immediately to its left and to its right. For example, $\binom{3}{1} = \binom{2}{0} + \binom{2}{1}$. Since the set of positive integers is closed under addition, all binomial coefficients are indeed positive integers.

3.2 Solutions

Problem 1938.1.

Prove that an integer n can be expressed as the sum of two squares if and only if $2n$ can be expressed as the sum of two squares.

Solution: Suppose that $x = a^2 + b^2$. Then $2x = (a + b)^2 + (a - b)^2$. Conversely, suppose $2x = c^2 + d^2$. Then c and d have the same parity. It follows that $\frac{c+d}{2}$ and $\frac{c-d}{2}$ are integers, and we have $x = \left(\frac{c+d}{2}\right)^2 + \left(\frac{c-d}{2}\right)^2$.

Problem 1942.2.

Let a, b, c and d be integers such that for all integers m and n, there exist integers x and y such that $ax + by = m$ and $cx + dy = n$. Prove that $ad - bc = \pm 1$.

First Solution: First suppose that $a = 0$. Then we can express any integer m in the form by, so that $b = \pm 1$, $cx = n - dy$ and c divides $n \mp dm$ for all m and n, and so $c = \pm 1$ and $ad - bc = \pm 1$. The argument is similar if any of b, c and d are 0.

If $abcd \neq 0$, let $\Delta = ad - bc$. Suppose that $\Delta = 0$. Then $\frac{c}{a} = \frac{d}{b}$. Let their common value be λ. Then $n = cx + dy = \lambda(ax + by) = \lambda m$. This means that $\frac{n}{m} = \lambda$ for any integers m and n. This is of course absurd. Hence $\Delta \neq 0$. We now solve $ax + by = m$ and $cx + dy = n$ for x and y. We have $x = \frac{dm - bn}{\Delta}$ and $y = \frac{an - cm}{\Delta}$. We are given that for any integers m and n, x and y are also integers. In particular, for $(m, n) = (1, 0)$, $x_1 = \frac{d}{\Delta}$ and $y_1 = -\frac{c}{\Delta}$ are integers, and for $(m, n) = (0, 1)$, $x_2 = -\frac{b}{\Delta}$ and $y_2 = \frac{a}{\Delta}$ are integers. It follows that $x_1 y_2 - x_2 y_1 = \frac{ad - bc}{\Delta^2} = \frac{1}{\Delta}$ is also an integer. The only integers whose reciprocals are also integers are ± 1. Since Δ is clearly an integer, we must have $\Delta = \pm 1$.

Remark: For another solution to this problem, see page 109 in Chapter 6.

Problem 1931.1.

Let p be a prime greater than 2. Prove that $\frac{2}{p}$ can be expressed in exactly one way in the form $\frac{1}{x} + \frac{1}{y}$ where x and y are positive integers with $x > y$.

First Solution: The equation may be rewritten as $2xy = p(x + y)$. By the Prime Divisibility Theorem, p divides at least one of 2, x and y. It clearly does not divide 2. Assume that p divides x. Hence $x = pa$ for some integer a. Then $2ay = pa + y$ or $(2a - 1)y = pa$. Now $\gcd(2a - 1, a) = 1$. By the Relatively Prime Divisibility Theorem, a divides y. Hence $y = ab$ for some integer b. Now $(2a - 1)b = p$. By the Fundamental Theorem of Arithmetic, one of $2a - 1$ and b must be p. If $b = p$, then $a = 1$ and $x = y = p$, but $x > y$. Hence $b = 1$ and $2a - 1 = p$, so that $x = \frac{p(p+1)}{2}$ and $y = \frac{p+1}{2}$, with $x > y$. If we had assumed that p divides y we would have arrived by the same reasoning at a solution where $x < y$. This is contrary to the hypothesis.

Second Solution: The equation may be rewritten as $4xy - 2px - 2py = 0$. Adding p^2 to both sides and factoring, we have $(2x - p)(2y - p) = p^2$. By the Fundamental Theorem of Arithmetic, there are only three possibilities. The first is $2x - p = p$ and $2y - p = p$. However, this implies $x = y = p$, which is to be rejected. Since $x > y$, we have that $2x - p = p^2$ and $2y - p = 1$. It follows that $x = \frac{p(p+1)}{2}$ and $y = \frac{p+1}{2}$.

Problem 1940.2.

Let m and n be distinct positive integers. Prove that $2^{2^m} + 1$ and $2^{2^n} + 1$ have no common divisor greater than 1.

Solution: Let $a_k = 2^{2^k}$. Then $a_k^2 = a_{k+1}$ and $a_{k+1} - 1 = (a_k - 1)(a_k + 1)$. Since $a_{k+1} - 1$ is divisible by $a_k - 1$, it is divisible by $a_\ell - 1$ for all $\ell \leq k$. We may assume that $m > n$. Then $a_n + 1$ divides $a_{n+1} - 1$, which in turn divides $a_m - 1$. It follows that $a_m + 1 = q(a_n + 1) + 2$ for some integer q. By the Distributivity Theorem, any common divisor of $a_m + 1$ and $a_n + 1$ must also be a divisor of 2. Since both $a_m + 1$ and $a_n + 1$ are odd, their greatest common divisor must be 1.

Remark: Integers of the form $2^{2^n} + 1$ are called **Fermat numbers**. For another approach to this problem, see page 57 in Chapter 4.

Problem 1931.2.

Let $a_1^2 + a_2^2 + a_3^2 + a_4^2 + a_5^2 = b^2$, where a_1, a_2, a_3, a_4, a_5 and b are integers. Prove that not all of these numbers can be odd.

Solution: We work in modulo 8 arithmetic. If k is an odd number, then $k \equiv 1, 3, 5$ or 7 and $k^2 \equiv 1$. If all of a_1, a_2, a_3, a_4, a_5 and b are odd, then $a_1^2 + a_2^2 + a_3^2 + a_4^2 + a_5^2 \equiv 5$ while $b^2 \equiv 1$. We have a contradiction.

Problem 1939.2.

Determine the highest power of 2 that divides $2^n!$.

First Solution: From $2^1! = 2, 2^2! = 24$ and $2^3! = 40320$, we see that the answers for the first three values of n are 1, 3 and 7 respectively. We conjecture that the answer in general is $2^n - 1$ and prove this by induction. The basis has already been established. We assume that for some $n \geq 3$, the highest power of 2 that divides $2^n!$ is $2^n - 1$. This means that the highest power of 2 that divides $(2^n - 1)!$ is $2^n - 1 - n$. Consider now $2^{n+1}!$. We line up the 2^{n+1} factors in two rows as follows:

1	2	3	$\cdots$	2^n
$2^n + 1$	$2^n + 2$	$2^n + 3$	$\cdots$	2^{n+1}

The highest power of 2 that divides the product of numbers in the first row is $2^n - 1$ by the induction hypothesis. For $1 \leq k < 2^n$, a power of 2 divides k if and only if it divides $2^n + k$. This follows from the Distributivity Theorem since such a power of 2 must divide 2^n. Hence in each column except the last, the highest power of 2 that divides the number in the first row is the same as that dividing the number in the second row. In the last column, the highest power of 2 dividing the number in the first row is 1 less than that dividing the number in the second row. Hence the highest power of 2 that divides $2^{n+1}!$ is $2(2^n - 1) + 1 = 2^{n+1} - 1$.

Second Solution: Each of the 2^n factors of $2^n!$ is given a number of cards equal to the highest power of 2 that divides that factor. We now collect these cards in n rounds, one card at a time from each factor that still has cards. In the first, since the odd factors do not have any cards, we take one from each of the even factors. Hence we can collect 2^{n-1} cards. In the second round, those factors that are divisible by 2 but not by 4 will no longer have cards. Thus we collect only from every fourth factor, for a total of 2^{n-2} cards. In each subsequent round, we collect half as many cards as the preceding one. In the last round, we only take from the last factor. It follows that the total number of cards collected is $2^{n-1} + 2^{n-2} + \cdots + 1 = 2^n - 1$. This is the highest power of 2 that divides $2^n!$.

Problem 1934.1.

Let n be a given positive integer and

$$A = \frac{1 \cdot 3 \cdot 5 \cdots (2n-1)}{2 \cdot 4 \cdot 6 \cdots 2n}.$$

Prove that at least one term of the sequence $A, 2A, 4A, 8A, \ldots, 2^k A, \ldots$ is an integer.

Solution: Multiply both the numerator and the denominator of A by $2 \cdot 4 \cdots 2n = 2^n \times n!$. Then

$$A = \frac{(2n)!}{(2^n n!)^2} = \frac{1}{2^{2n}} \binom{2n}{n}.$$

Since the binomial coefficient is an integer, $2^k A$ is an integer for all $k \geq 2n$.

Problem 1932.1.

Let a, b and n be positive integers such that b is divisible by a^n. Prove that $(a+1)^b - 1$ is divisible by a^{n+1}.

First Solution: Since

$$(a+1)^b - 1 = \left[(a+1) - 1\right]\left[(a+1)^{b-1} + (a+1)^{b-2} + \cdots + (a+1) + 1\right]$$

is divisible by a, the result is also true for $n = 0$. We now use induction on n. Suppose the result is true for some $n \geq 0$. Let c be a positive integer such that a^{n+1} divides c. Let $c = ab$. Then a^n divides b. By the induction hypothesis, $(a+1)^b - 1$ is divisible by a^{n+1}. Note that $(a+1)^c - 1 = (a+1)^{ab} - 1$ expands into

$$\left[(a+1)^b - 1\right]\left[(a+1)^{(a-1)b} + (a+1)^{(a-2)b} + \cdots + (a+1)^b + 1\right].$$

It remains to be shown that a divides the second factor. There are a terms in that factor, which may be rewritten as

$$\left[(a+1)^{(a-1)b} - 1\right] + \left[(a+1)^{(a-2)b} - 1\right] + \cdots + \left[(a+1)^b - 1\right] + a.$$

The last term is obviously divisible by a, and the argument for the case $n = 0$ shows that each of the other terms is divisible by a also. By the Distributivity Theorem, the second factor is indeed divisible by a. This completes the induction argument.

Second Solution: We have $(a+1)^b-1 = \binom{b}{1}a+\binom{b}{2}a^2+\cdots+\binom{b}{b}a^b$ by the Binomial Theorem. We now prove that a^{n+1} divides $\binom{b}{k}a^k$ for $1 \leq k \leq b$. Note that we have $k\binom{b}{k} = b\binom{b-1}{k-1}$. Let d be the greatest common divisor of k and b. Then $\frac{k}{d}$ and $\frac{b}{d}$ are relatively prime to each other. Now $\frac{b}{d}$ divides $\frac{k}{d}\binom{b}{k}$. By the Relatively Prime Divisibility Theorem, $\frac{b}{d}$ divides $\binom{b}{k}$. It remains to be shown that a^{n+1-k} divides $\frac{b}{d}$. For any prime divisor p of a, let the highest power of p dividing a be $\alpha \geq 1$. Since a^n divides b, the highest power of p dividing b is at least $n\alpha$. Since $d \leq k < 2^k \leq p^k$, the highest power of p dividing d is at most $k-1$. It follows that the highest power of p dividing $\frac{b}{d}$ is at least $n\alpha - (k-1)$. On the other hand, the highest power of p dividing a^{n+1-k} is $\alpha(n+1-k) \leq n\alpha - (k-1)$. Since p is an arbitrary prime divisor of a, we have a^{n+1-k} dividing $\frac{b}{d}$ as desired.

Problem 1936.3.

Let a be any positive integer. Prove that there exists a unique pair of positive integers x and y such that $x + \frac{1}{2}(x+y-1)(x+y-2) = a$.

First Solution: Observe that the given equation may be rewritten as $x + \binom{x+y-1}{2} = a$. We first construct a solution for any given a. Let n be the largest integer such that $a > \binom{n}{2}$. If $a = 1$, then $n = 1$. Let $x = a - \binom{n}{2}$ and $y = n - x + 1$. By Pascal's Formula, $x \leq \binom{n+1}{2} - \binom{n}{2} = \binom{n}{1} = n$, so that $y \geq 1$. We have $x + \frac{1}{2}(x+y-1)(x+y-2) = a$. We now prove uniqueness. Suppose that x and y are positive integers satisfying this equation. Let n be as before. If $n < x+y-1$, then $x = a - \binom{x+y-1}{2} \leq \binom{x+y-1}{2} - \binom{x+y-1}{2} = 0$. If $n > x+y-1$, then

$$x = a - \binom{x+y-1}{2} > \binom{n}{2} - \binom{x+y-1}{2}$$
$$\geq \binom{x+y}{2} - \binom{x+y-1}{2}$$
$$= \binom{x+y-1}{1}$$
$$= x+y-1$$

by Pascal's Formula. This implies that $y < 1$, which is a contradiction. It follows that we must have $x+y-1 = n$. Now $x = a - \binom{n}{2}$ is uniquely determined. Hence $y = n - x + 1$ is also uniquely determined.

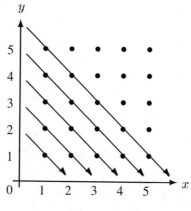

FIGURE 3.2.2

Second Solution: Each pair (x, y) of positive integers may be associated with the point in the first quadrant with coordinates x and y. Order these points along diagonals downward to the right, starting with the one nearest the origin. The first few points in this sequence are $(1,1)$, $(1,2)$, $(2,1)$, $(1,3)$, $(2,2)$ and $(3,1)$. Figure 3.2.2 illustrates the overall scheme.

For any positive integer k, the number of points (x, y) with $x + y = k$ is $k - 1$. Count the number of points in the sequence preceding a particular point (x, y). The number of such points in earlier diagonals is $1 + 2 + \cdots + (x + y - 2) = \frac{1}{2}(x + y - 1)(x + y - 2)$. The number of such points in the same diagonal is $x - 1$. Hence the point (x, y) is in the ath position in the sequence, where $a = x + \frac{1}{2}(x + y - 1)(x + y - 2)$. Since each point (x, y) has a unique position a in the sequence, each positive integer a corresponds to a unique pair of positive integers (x, y).

4

Algebra Problems

Problem 1933.1.

Let a, b, c and d be real numbers such that $a^2 + b^2 = c^2 + d^2 = 1$ and $ac + bd = 0$. Determine the value of $ab + cd$.

Problem 1936.1.

Prove that for all positive integers n,

$$\frac{1}{1 \cdot 2} + \frac{1}{3 \cdot 4} + \cdots + \frac{1}{(2n-1)2n} = \frac{1}{n+1} + \frac{1}{n+2} + \cdots + \frac{1}{2n}.$$

Problem 1941.1.

Prove that

$$(1+x)(1+x^2)(1+x^4)(1+x^8) \cdots (1+x^{2^{k-1}}) = 1+x+x^2+x^3+\cdots+x^{2^k-1}.$$

Problem 1943.3.

Let $a < b < c < d$ be real numbers and $\langle x, y, z, t \rangle$ be any permutation of a, b, c and d. What are the maximum and minimum values of the expression $(x - y)^2 + (y - z)^2 + (z - t)^2 + (t - x)^2$?

Problem 1938.2.

Prove that for all integers $n > 1$, $\dfrac{1}{n} + \dfrac{1}{n+1} + \cdots + \dfrac{1}{n^2-1} + \dfrac{1}{n^2} > 1$.

Problem 1937.1.

Let n be a positive integer. Prove that $a_1! \, a_2! \cdots a_n! < k!$, where k is an integer that is greater than the sum of the positive integers $a_1, a_2, \ldots, a_n$.

Problem 1935.1.

Let n be a positive integer. Prove that $\dfrac{a_1}{b_1} + \dfrac{a_2}{b_2} + \cdots + \dfrac{a_n}{b_n} \geq n$, where $\langle b_1, b_2, \ldots, b_n \rangle$ is any permutation of the positive real numbers $a_1, a_2, \ldots, a_n$.

Problem 1939.1.

Let a_1, a_2, b_1, b_2, c_1 and c_2 be real numbers for which $a_1 a_2 > 0$, $a_1 c_1 \geq b_1^2$ and $a_2 c_2 \geq b_2^2$. Prove that $(a_1 + a_2)(c_1 + c_2) \geq (b_1 + b_2)^2$.

Problem 1929.2.

Let $k \leq n$ be positive integers and x be a real number with $0 \leq x < \frac{1}{n}$. Prove that $\binom{n}{0} - \binom{n}{1}x + \binom{n}{2}x^2 - \cdots + (-1)^k \binom{n}{k}x^k > 0$.

4.1 Discussion

4.1.1 Inequalities

Of all the topics commonly featured in competitions, Algebra is the most familiar to high school students. Here, exercises are plentiful, but good problems are harder to come by. Many competition problems in this topic are inequalities, a subject that is not thoroughly covered in the standard curriculum.

We have already encountered inequalities informally in Chapter 2. Let us give a more rigorous treatment here. A set of numbers S is said to be **ordered** if it contains a subset P of elements with the following three properties:

1. **Trichotomy:** $a \in S \implies a \in P, a = 0$ or $-a \in P$.
2. **Additivity:** $a, b \in P \implies a + b \in P$.
3. **Multiplicativity:** $a, b \in P \implies ab \in P$.

The elements in P are usually called the *positive* elements of S. By the Trichotomy property, 0 is not positive. An important example of an ordered number system is the real number system $\mathcal{R}$. The familiar binary relation $<$ is defined as follow: $x < y$ if and only if $y - x$ is positive. By $x \leq y$ is meant $x < y$ or $x = y$. Sometimes, the symbols $>$ and $\geq$ are used, with obvious meanings.

The following properties are easily derived from those of order.

1. **Reflexivity:** $x \in \mathcal{R} \Longrightarrow x \leq x$.
2. **Antisymmetry:** $x \leq y$ and $y \leq x \Longrightarrow x = y$.
3. **Transitivity:** $x \leq y$ and $y \leq z \Longrightarrow x \leq z$.
4. **Trichotomy:** $x \in \mathcal{R} \Longrightarrow x < 0, x = 0$ or $0 < x$.
5. **Additivity:** $w \leq x$ and $y \leq z \Longrightarrow w + y \leq x + z$.
6. **Multiplicativity:** $x \leq y$ and $0 \leq z \Longrightarrow xz \leq yz$,
   while $x \leq y$ and $z \leq 0 \Longrightarrow yz \leq xz$.

The first three properties show that "less than or equal to" is an order relation like divisibility. The fourth property shows that it is a **total order** relation, which divisibility is not. Note that $w \leq x$ and $y \leq z$ do not in general imply that $wy \leq xz$. For example, if x is any negative real number, then $x < 0$ but $x^2 = (-x)^2 > 0$. In fact, $x^2 \geq 0$ for any real number x.

We mentioned in Chapter 2 that we will define more mean functions here. The **geometric mean** of the *positive* real numbers $x_1, x_2, \ldots, x_n$ is defined to be $M_0(x_1, x_2, \ldots, x_n) = \sqrt[n]{x_1 x_2 \cdots x_n}$. It is easy to see that

$$M_{-\infty}(x_1, x_2, \ldots, x_n) \leq M_0(x_1, x_2, \ldots, x_n) \leq M_\infty(x_1, x_2, \ldots, x_n).$$

Perhaps the most important and best known inequality is the **Arithmetic-Geometric Mean Inequality**, which states that for any *positive* real numbers $x_1, x_2, \ldots, x_n$, we have

$$M_1(x_1, x_2, \ldots, x_n) \geq M_0(x_1, x_2, \ldots, x_n).$$

Equality holds if and only if $x_1 = x_2 = \cdots = x_n$.

We shall prove this result using an unusual form of induction. For $n = 2$, we use the fact that $(\sqrt{x_1} - \sqrt{x_2})^2 \geq 0$. Expansion yields $x_1 + x_2 \geq 2\sqrt{x_1 x_2}$, which yields the desired result upon division by 2. Equality holds if and only if $\sqrt{x_1} = \sqrt{x_2}$, and since the numbers are positive, this is equivalent to $x_1 = x_2$.

Suppose the inequality holds for some $n \geq 2$. Instead of considering the next case $n+1$, we jump forward to the case $2n$. Using both the induction hypothesis as well as the established case $n = 2$, we have

$$\frac{x_1 + x_2 + \cdots + x_{2n}}{2n} = \frac{\dfrac{x_1 + x_2 + \cdots + x_n}{n} + \dfrac{x_{n+1} + x_{n+2} + \cdots + x_{2n}}{n}}{2}$$

$$\geq \frac{\sqrt[n]{x_1 x_2 \cdots x_n} + \sqrt[n]{x_{n+1} x_{n+2} \cdots x_{2n}}}{2}$$

$$\geq \sqrt[2n]{x_1 x_2 \cdots x_{2n}}.$$

In order for equality to hold, we must have $x_1 = x_2 = \cdots = x_n$, $x_{n+1} = x_{n+2} = \cdots = x_{2n}$ and $\sqrt[n]{x_1 x_2 \cdots x_n} = \sqrt[n]{x_{n+1} x_{n+1} \cdots x_{2n}}$. This is equivalent to $x_1 = x_2 = \cdots = x_{2n}$.

This doubling process allows us to prove the result whenever n is a power of 2, and this sequence is not bounded above. To complete the proof of the Arithmetic-Geometric Mean Inequality, we come back to fill in the missing gaps. More specifically, we show that if the result holds for some $n \geq 4$, then it also holds for the preceding case $n - 1$.

Let $x_1, x_2, \ldots, x_{n-1}$ be any positive real numbers. Denote their arithmetic mean by x. Then the arithmetic mean of $x_1, x_2, \ldots, x_{n-1}$ and x is also equal to x. It follows from the induction hypothesis that $x \geq \sqrt[n]{x_1 x_2 \cdots x_{n-1} x}$, that is $x^{n-1} \geq x_1 x_2 \cdots x_{n-1}$. Taking the $(n-1)$st root yields the desired result. Equality holds if and only if $x_1 = x_2 = \cdots = x_{n-1}$.

Let $x_1, x_2, \ldots, x_n$ be positive real numbers. For any non-zero real number t, we define the tth power mean function to be

$$M_t(x_1, x_2, \ldots, x_n) = \left(\frac{x_1^t + x_2^t + \cdots + x_n^t}{n} \right)^{\frac{1}{t}}.$$

The **Power Means Inequality** states that $M_s \geq M_t$ if $s > t$, with equality if and only if $x_1 = x_2 = \cdots = x_n$. This is a powerful result, but we will not prove it beyond the special cases that we have already considered.

4.1.2   The Rearrangement Inequality

Let $a_1 \leq a_2 \leq \cdots \leq a_n$ and $b_1 \leq b_2 \leq \cdots \leq b_n$ be real numbers. Let $\langle c_1, c_2, \ldots, c_n \rangle$ be any permutation of $b_1, b_2, \ldots, b_n$. The **Rearrangement Inequality** states that

$$a_1 b_n + a_2 b_{n-1} + \cdots + a_n b_1 \leq a_1 c_1 + a_2 c_2 + \cdots + a_n c_n$$

$$\leq a_1 b_1 + a_2 b_2 + \cdots + a_n b_n.$$

We first prove that

$$a_1 c_1 + a_2 c_2 + \cdots + a_n c_n \leq a_1 b_1 + a_2 b_2 + \cdots + a_n b_n.$$

If $b_i = c_i$ for all i, these expressions are obviously equal. Otherwise, consider the smallest $i \geq 1$ where $b_i \neq c_i$. Now b_i is the minimum of the set $B = \{b_i, b_{i+1}, \ldots, b_n\}$, and since the set $C = \{c_i, c_{i+1}, \ldots, c_n\}$ is a permutation of the set B, then b_i is also the minimum of C. Since $c_i \neq b_i$, then c_i is not the minimum and it follows that $c_i > b_i$. However, b_i occurs somewhere in C, implying $b_i = c_j$ for some $j \geq i$. Hence $c_i > c_j$.

Now we interchange c_i and c_j and claim that

$$a_1c_1 + a_2c_2 + \cdots + a_nc_n \leq a_1c_1 + \cdots + a_ic_j + \cdots + a_jc_i + \cdots + a_nc_n.$$

We have only to compare $a_ic_i + a_jc_j$ with $a_ic_j + a_jc_i$. The latter cannot be less than the former since

$$(a_ic_j + a_jc_i) - (a_ic_i + a_jc_j) = (a_j - a_i)(c_i - c_j) \geq 0.$$

Note that since $c_i > c_j$, equality holds only if $a_i = a_j$. Repeating this switching process, we eventually convert $a_1c_1 + a_2c_2 + \cdots + a_nc_n$ into $a_1b_1 + a_2b_2 + \cdots + a_nb_n$. Since the sum does not decrease at any point, we have the desired result.

When will equality hold? It is tempting to say that this happens if and only if $c_i = b_i$ for $1 \leq i \leq n$, but only with the convention that $c_i \leq c_{i+1}$ whenever $a_i = a_{i+1}$. Without this convention, consider the extreme case where all the a's are identical. Then equality holds no matter how the b's are permuted.

We could use the same approach as before to prove the other half of the Rearrangement Inequality, but choose to present instead a proof by induction on n. For $n = 1$, we have $a_1b_1 \leq a_1b_1$. We now prove that

$$a_1b_{n+1} + a_2b_n + \cdots + a_{n+1}b_1 \leq a_1c_1 + a_2c_2 + \cdots + a_{n+1}c_{n+1},$$

assuming as the induction hypothesis that

$$a_1b_{n+1} + a_2b_n + \cdots + a_nb_2 \leq a_1d_1 + a_2d_2 + \cdots + a_nd_n$$

where $\langle d_1, d_2, \ldots, d_n \rangle$ is any permutation of $b_2, b_3, \ldots, b_{n+1}$.

Compare the terms $a_{n+1}b_1$ and $a_{n+1}c_{n+1}$. If $c_{n+1} = b_1$, they can be cancelled, and the desired result follows from the induction hypothesis with $d_i = c_i$ for $1 \leq i \leq n$. Otherwise, we must have $c_{n+1} > b_1$, and $b_1 = c_k$ for some $k \leq n$. Then we have

$$(a_kc_k + a_{n+1}c_{n+1}) - (a_kc_{n+1} + a_{n+1}c_k) = (a_{n+1} - a_k)(c_{n+1} - c_k) \geq 0.$$

By the induction hypothesis with $d_i = c_i$ for $1 \leq i \leq n$ except $d_k = c_{n+1}$, we have $c_{n+1} > c_k$ and

$$a_1b_{n+1} + a_2b_n + \cdots + a_{n+1}b_1$$
$$\leq a_1c_1 + \cdots + a_kc_{n+1} + \cdots + a_nc_n + a_{n+1}c_k$$
$$\leq a_1c_1 + a_2c_2 + \cdots + a_{n+1}c_{n+1}.$$

Equality holds if and only if $c_i = b_{n-i+1}$ for $1 \le i \le n$, with the convention that $c_i \ge c_{i+1}$ whenever $a_i = a_{i+1}$.

Note that nowhere in the proof did we need the a_i and b_i to be positive. This is quite often not the case with other inequalities. For example, even the statement of the Arithmetic-Geometric Mean Inequality may not make sense without this assumption. Thus the Rearrangement Inequality is a surprisingly strong result. In particular, it can be used to derive a number of classical inequalities.

As an example, we prove an important result known as **Cauchy's Inequality**. Let $a_1, a_2, \ldots, a_n$, $b_1, b_2, \ldots, b_n$ be real numbers. Then

$$(a_1 b_1 + a_2 b_2 + \cdots + a_n b_n)^2 \le (a_1^2 + a_2^2 + \cdots + a_n^2)(b_1^2 + b_2^2 + \cdots + b_n^2),$$

with equality if and only if for some constant $k, a_i = k b_i$ for $1 \le i \le n$ or $b_i = k a_i$ for $1 \le i \le n$.

If $a_1 = a_2 = \cdots = a_n = 0$ or $b_1 = b_2 = \cdots = b_n = 0$, the result is trivial. Otherwise we define $S = \sqrt{a_1^2 + a_2^2 + \cdots + a_n^2}$ and $T = \sqrt{b_1^2 + b_2^2 + \cdots + b_n^2}$. Since both are non-zero, we may let $x_i = \frac{a_i}{S}$ and $x_{n+i} = \frac{b_i}{T}$ for $1 \le i \le n$.

By the Rearrangement Inequality,

$$2 = \frac{a_1^2 + a_2^2 + \cdots + a_n^2}{S^2} + \frac{b_1^2 + b_2^2 + \cdots + b_n^2}{T^2}$$

$$= x_1^2 + x_2^2 + \cdots + x_{2n}^2$$

$$\ge x_1 x_{n+1} + x_2 x_{n+2} + \cdots + x_n x_{2n} + x_{n+1} x_1 + x_{n+2} x_2 + \cdots + x_{2n} x_n$$

$$= \frac{2(a_1 b_1 + a_2 b_2 + \cdots + a_n b_n)}{ST},$$

which is equivalent to the desired result. Equality holds if and only if $x_i = x_{n+i}$ for $1 \le i \le n$, or $a_i T = b_i S$ for $1 \le i \le n$.

It should be mentioned that there is a dual form of the Rearrangement Inequality, discovered recently by **Robert Geretschläger** of Graz, Austria. See his paper *Deriving Some Problems in Inequalities from an Accidental Generalization*, Mathematics Competitions **12** (1999) #1, 27–37.

Let $a_1 \le a_2 \le \cdots \le a_n$ and $b_1 \le b_2 \le \cdots \le b_n$ be positive real numbers. Let $\langle c_1, c_2, \ldots, c_n \rangle$ be any permutation of $b_1, b_2, \ldots, b_n$. Then

$$(a_1 + b_n)(a_2 + b_{n-1}) \cdots (a_n + b_1)$$

$$\ge (a_1 + c_1)(a_2 + c_2) \cdots (a_n + c_n)$$

$$\ge (a_1 + b_1)(a_2 + b_2) \cdots (a_n + b_n).$$

It can be proved in the same way as the Rearrangement Inequality.

For $n = 2$, we only have to show that

$$(a_1 + b_2)(a_2 + b_1) \geq (a_1 + b_1)(a_2 + b_2).$$

Subtracting the second product from the first, we obtain

$$(a_2 - a_1)(b_2 - b_1) \geq 0.$$

This holds whether the four numbers are positive or not. However, the general result is false without this assumption. For instance, suppose $n = 3$, $a_1 = 1$, $a_2 = 2$, $a_3 = 3$, $b_1 = -2$, $b_2 = 1$ and $b_3 = 2$. Then

$$(a_1 + b_1)(a_2 + b_2)(a_3 + b_3) = -15.$$

However, if $c_1 = b_1$, $c_2 = b_3$ and $c_3 = b_2$, then

$$(a_1 + c_1)(a_2 + c_2)(a_3 + c_3) = -16.$$

4.2   Solutions

Problem 1933.1.

Let a, b, c and d be real numbers such that $a^2 + b^2 = c^2 + d^2 = 1$ and $ac + bd = 0$. Determine the value of $ab + cd$.

First Solution: Since $a^2 + b^2 = 1$, a and b are not both 0. We may assume that $a \neq 0$. From $ac + bd = 0$, we have $c = -\frac{bd}{a}$. Substituting into $c^2 + d^2 = 1$, we have

$$1 = \frac{b^2 d^2}{a^2} + d^2 = \frac{(a^2 + b^2)d^2}{a^2} = \frac{d^2}{a^2}.$$

It follows that $a^2 = d^2$. Hence

$$ab + cd = ab - \frac{bd^2}{a} = \frac{(a^2 - d^2)b}{a} = 0.$$

Second Solution: By symmetry, we could have replaced the hypothesis $ac + bd = 0$ with $ad + bc = 0$. Thus we only need to know either condition holds, so that we can weaken the hypothesis to $(ac + bd)(ad + bc) = 0$. Expanding the left side, we have $ab(c^2 + d^2) + cd(a^2 + b^2)$. In view of $a^2 + b^2 = 1 = c^2 + d^2$, we have $ab + cd = 0$ as desired.

Remark: For other solutions to this problem, see pages 116 and 117 in Chapter 6.

Problem 1936.1.

Prove that for all positive integers n,

$$\frac{1}{1\cdot 2} + \frac{1}{3\cdot 4} + \cdots + \frac{1}{(2n-1)2n} = \frac{1}{n+1} + \frac{1}{n+2} + \cdots + \frac{1}{2n}.$$

First Solution: We have

$$\frac{1}{1\cdot 2} + \frac{1}{3\cdot 4} + \cdots + \frac{1}{(2n-1)2n}$$

$$= \left(1 - \frac{1}{2}\right) + \left(\frac{1}{3} - \frac{1}{4}\right) + \cdots + \left(\frac{1}{2n-1} - \frac{1}{2n}\right)$$

$$= 1 + \frac{1}{2} + \frac{1}{3} + \frac{1}{4} + \cdots + \frac{1}{2n-1} + \frac{1}{2n} - 2\left(\frac{1}{2} + \frac{1}{4} + \cdots + \frac{1}{2n}\right)$$

$$= 1 + \frac{1}{2} + \cdots + \frac{1}{2n} - \left(1 + \frac{1}{2} + \cdots + \frac{1}{n}\right)$$

$$= \frac{1}{n+1} + \frac{1}{n+2} + \cdots + \frac{1}{2n}.$$

Second Solution: Denote the left side by $S(n)$ and the right side by $T(n)$. We prove by induction that $S(n) = T(n)$ for all $n \geq 1$. For a basis we note that $S(1) = \frac{1}{1\cdot 2} = \frac{1}{1+1} = T(1)$. Suppose $S(n) = T(n)$ for some $n \geq 1$. Then

$$T(n+1) - T(n) = \frac{1}{2n+1} + \frac{1}{2n+2} - \frac{1}{n+1}$$

$$= \frac{1}{(2n+1)(2n+2)}$$

$$= S(n+1) - S(n).$$

By the induction hypothesis, $S(n+1) = T(n+1)$. This completes the inductive argument.

Problem 1941.1.

Prove that

$$(1+x)(1+x^2)(1+x^4)(1+x^8)\cdots(1+x^{2^{k-1}})$$

$$= 1 + x + x^2 + x^3 + \cdots + x^{2^k - 1}.$$

First Solution: We use induction on k. For $k = 1$, both sides are equal to $1 + x$. Suppose that the identity holds for some $k \geq 1$. By the induction hypothesis,

$$(1+x)(1+x^2)\cdots(1+x^{2^{k-1}})(1+x^{2^k})$$
$$= (1+x+x^2+\cdots+x^{2^k-1})(1+x^{2^k})$$
$$= 1+x+\cdots+x^{2^k-1}+x^{2^k}+x^{2^k+1}+\cdots x^{2^{k+1}-1}.$$

Second Solution: If $x = 1$, both sides are equal to 2^k and the identity holds. If $x \neq 1$, the identity still holds since

$$(x-1)(x+1)(x^2+1)(x^4+1)\cdots(x^{2^{k-1}}+1)$$
$$= (x^2-1)(x^2+1)(x^4+1)\cdots(x^{2^{k-1}}+1)$$
$$= (x^4-1)(x^4+1)\cdots(x^{2^{k-1}}+1)$$
$$= \cdots$$
$$= x^{2^k}-1$$
$$= (x-1)(1+x+x^2+\cdots+x^{2^k-1}).$$

Remark: We can now give an alternative solution to **Problem 1940.2**. Let F_i denote the ith Fermat number $2^{2^i}+1$. Setting $x = 2$ in the identity of this problem and adding 2 to both sides, the right side becomes $1+2+2^2+\cdots+2^{2^k-1}+2 = 2^{2^k}+1 = F_k$ while the left side becomes $F_0 F_1 F_2 \cdots F_{k-1}+2$. When $i \leq k$, a number that divides both F_i and F_k must also divide 2. Since both F_i and F_k are odd, their greatest common divisor must be 1.

Problem 1943.3.

Let $a < b < c < d$ be real numbers and $\langle x, y, z, t \rangle$ be any permutation of a, b, c and d. What are the maximum and minimum values of the expression $(x-y)^2 + (y-z)^2 + (z-t)^2 + (t-x)^2$?

First Solution: Let us denote the given expression by $f(x, y, z, t)$. By cyclic symmetry, we may take $x = a$. Clearly we have $f(a, b, c, d) = f(a, d, c, b)$, $f(a, b, d, c) = f(a, c, d, b)$ and $f(a, c, b, d) = f(a, d, b, c)$. We now prove that these three values are distinct. Indeed,

$$f(a, b, c, d) - f(a, b, d, c) = (b-c)^2 + (d-a)^2 - (b-d)^2 - (c-a)^2$$
$$= 2(b-a)(d-c)$$
$$> 0.$$

Similarly, $f(a, c, b, d) - f(a, b, c, d) = 2(c-b)(d-a) > 0$. Hence $f(a, c, b, d) > f(a, b, c, d) > f(a, b, d, c)$.

Second Solution: Let us denote the given expression by $f(x, y, z, t)$. Then the value of $f(x, y, z, t) - 2(xz + yt)$ is

$$(x-y)^2 + (x-z)^2 + (x-t)^2 + (y-z)^2 + (y-t)^2 + (z-t)^2 - x^2 - y^2 - z^2 - t^2.$$

This does not depend on which permutation of a, b, c and d is used. Now $xz + yt$ has at most three distinct values, namely, $ab + cd$, $ac + bd$ and $ad + bc$. Since $ab + cd - (ac + bd) = (d - a)(c - b) > 0$ and $ac + bd - (ad + bc) = (b - a)(d - c) > 0$, $f(a, c, b, d)$ is the maximum value and $f(a, b, d, c)$ is the minimum value.

Problem 1938.2.

Prove that for all integers $n > 1$, $\dfrac{1}{n} + \dfrac{1}{n+1} + \cdots + \dfrac{1}{n^2 - 1} + \dfrac{1}{n^2} > 1$.

Solution: We have

$$\frac{1}{n} + \frac{1}{n+1} + \cdots + \frac{1}{n^2 - 1} + \frac{1}{n^2} > \frac{1}{n} + \frac{1}{n^2} + \cdots + \frac{1}{n^2} + \frac{1}{n^2}$$

$$= \frac{1}{n} + \frac{n^2 - n}{n^2}$$

$$= \frac{1}{n} + \frac{n - 1}{n}$$

$$= 1.$$

Problem 1937.1.

Let n be a positive integer. Prove that $a_1! \, a_2! \cdots a_n! < k!$, where k is an integer which is greater than the sum of the positive integers $a_1, a_2, \ldots, a_n$.

Solution: Note that $a_1! \, a_2! \cdots a_n!$ and $(a_1 + a_2 + \cdots + a_n)!$ have the same number of factors. The first a_1 factors are the same in each expression. The next a_2 factors of the first expression are $1, 2, \ldots, a_2$ while the corresponding factors in the second expression are each greater by a_1. Similarly, each subsequent factor in the second expression exceeds the corresponding factor in the first expression. It follows that $a_1! \, a_2! \cdots a_n! \leq (a_1 + a_2 + \cdots + a_n)!$. Since $k! > (a_1 + a_2 + \cdots + a_n)!$, the desired result follows.

Problem 1935.1.

Let n be a positive integer. Prove that

$$\frac{a_1}{b_1} + \frac{a_2}{b_2} + \cdots + \frac{a_n}{b_n} \geq n,$$

where $\langle b_1, b_2, \ldots, b_n \rangle$ is any permutation of the positive real numbers $a_1, a_2, \ldots, a_n$.

First Solution: If $a_1 \le a_2 \le \cdots \le a_n$, then $\frac{1}{a_1} \ge \frac{1}{a_2} \ge \cdots \ge \frac{1}{a_n}$. By the Rearrangement Inequality,

$$\frac{a_1}{b_1} + \frac{a_2}{b_2} + \cdots + \frac{a_n}{b_n} \ge \frac{a_1}{a_1} + \frac{a_2}{a_2} + \cdots + \frac{a_n}{a_n} = n.$$

Second Solution: By the Arithmetic-Geometric Mean Inequality, we have

$$\frac{1}{n}\left(\frac{a_1}{b_1} + \frac{a_2}{b_2} + \cdots + \frac{a_n}{b_n} \right) \ge \sqrt[n]{\frac{a_1}{b_1} \cdot \frac{a_2}{b_2} \cdots \frac{a_n}{b_n}} = 1.$$

This is equivalent to the desired inequality.

Third Solution: We use induction on n. For $n = 1$, we must have $a_1 = b_1$ so that we certainly have $\frac{a_1}{b_1} \ge 1$. Suppose the inequality holds for some $n \ge 1$. Consider now the sum $S = \frac{a_1}{b_1} + \frac{a_2}{b_2} + \cdots + \frac{a_{n+1}}{b_{n+1}}$. In the special case where $a_i = b_i$ for some i, we have $\frac{a_i}{b_i} = 1$. By the induction hypothesis, the sum of the other n fractions in S is at least n. Hence $S \ge n+1$ as desired. In the general case where $a_i \ne b_i$ for $1 \le i \le n+1$, let a_1 be one of the largest a_i so that $a_1 \ge a_i$ for $2 \le i \le n + 1$. Now $a_1 = b_j$ for some j, and $\frac{1}{b_j} \le \frac{1}{b_i}$ for $i \ne j$. Denote by S' the sum obtained from S by switching b_1 and b_j. By the special case considered earlier, $S' \ge n + 1$. The desired inequality follows from

$$S - S' = \frac{a_1}{b_1} + \frac{a_j}{b_j} - \frac{a_1}{b_j} - \frac{a_j}{b_1}$$

$$= (a_1 - a_j)\left(\frac{1}{b_1} - \frac{1}{b_j} \right)$$

$$\ge 0.$$

Remark: We can now give an alternative proof of the Arithmetic-Geometric Mean Inequality, since the first solution uses only the Rearrangement Inequality. Let

$$G = \sqrt[n]{x_1 x_2 \cdots x_n},$$

$a_1 = \frac{x_1}{G}$, $a_2 = \frac{x_1 x_2}{G^2}, \ldots, a_n = \frac{x_1 x_2 \cdots x_n}{G^n} = 1$. Then

$$n \le \frac{a_1}{a_n} + \frac{a_2}{a_1} + \cdots + \frac{a_n}{a_{n-1}} = \frac{x_1}{G} + \frac{x_2}{G} + \cdots + \frac{x_n}{G},$$

which is equivalent to $\frac{x_1 + x_2 + \cdots + x_n}{n} \ge G$. Equality holds if and only if $a_1 = a_2 = \cdots = a_n$, or $x_1 = x_2 = \cdots = x_n$.

Problem 1939.1.

Let a_1, a_2, b_1, b_2, c_1 and c_2 be real numbers for which $a_1 a_2 > 0$, $a_1 c_1 \geq b_1^2$ and $a_2 c_2 \geq b_2^2$. Prove that $(a_1 + a_2)(c_1 + c_2) \geq (b_1 + b_2)^2$.

First Solution: In the lower bounds b_1^2 and b_2^2, it does not matter what signs b_1 and b_2 have. The lower bound $(b_1 + b_2)^2$ is larger if b_1 and b_2 have the same sign, so assume that both are positive. Suppose $a_1 > 0$. Then $a_2 > 0$, $c_1 \geq 0$ and $c_2 \geq 0$. We have $a_1 c_2 + a_2 c_1 \geq 2\sqrt{a_1 c_2 a_2 c_1} \geq 2 b_1 b_2$ by the Arithmetic-Geometric Mean Inequality. Hence

$$(a_1 + a_2)(c_1 + c_2) = a_1 c_1 + (a_1 c_2 + a_2 c_1) + a_2 c_2$$
$$\geq b_1^2 + 2 b_1 b_2 + b_2^2 = (b_1 + b_2)^2.$$

If $a_1 < 0$, then $a_2 < 0$, $c_1 \leq 0$ and $c_2 \leq 0$. The same argument applies.

Second Solution: Again we assume that $b_1 \geq 0$ and $b_2 \geq 0$. By Cauchy's Inequality,

$$(a_1 + a_2)(c_1 + c_2) = \left[\left(\sqrt{|a_1|} \right)^2 + \left(\sqrt{|a_2|} \right)^2 \right] \left[\left(\sqrt{|c_1|} \right)^2 + \left(\sqrt{|c_2|} \right)^2 \right]$$
$$\geq \left(\sqrt{|a_1|}\sqrt{|c_1|} + \sqrt{|a_2|}\sqrt{|c_2|} \right)^2$$
$$= \left(\sqrt{|a_1 c_1|} + \sqrt{|a_2 c_2|} \right)^2$$
$$\geq \left(\sqrt{b_1^2} + \sqrt{b_2^2} \right)^2 = (b_1 + b_2)^2.$$

Problem 1929.2.

Let $k \leq n$ be positive integers and x be a real number with $0 \leq x < \frac{1}{n}$. Prove that

$$\binom{n}{0} - \binom{n}{1} x + \binom{n}{2} x^2 - \cdots + (-1)^k \binom{n}{k} x^k > 0.$$

Solution: We partition the sum into pairs of consecutive terms, with possibly an extra term at the end. If there is an extra term at the end, the power in that term is even, and the term is positive. We now prove that the combined value of each of the pairs of terms is positive. Since $0 \leq x \leq \frac{1}{n}$, we have

$$\binom{n}{2i} x^{2i} - \binom{n}{2i+1} x^{2i+1} = \left(1 - \frac{(n-2i)x}{2i+1} \right) \binom{n}{2i} x^{2i}$$
$$= \frac{2i(1+x) + (1-nx)}{2i+1} \binom{n}{2i} x^{2i} > 0.$$

5
Geometry Problems
Part I

Problem 1931.3.

Let A and B be two given points, distance 1 apart. Determine a point P on the line AB such that $\frac{1}{1+AP} + \frac{1}{1+BP}$ is a maximum.

Problem 1942.1.

Prove that in any triangle, at most one side can be shorter than the altitude from the opposite vertex.

Problem 1937.3.

Let n be a positive integer. Let P, Q, A_1, $A_2, \ldots, A_n$ be distinct points such that A_1, $A_2, \ldots, A_n$ are not collinear. Suppose that $PA_1 + PA_2 + \cdots + PA_n$ and $QA_1 + QA_2 + \cdots + QA_n$ have a common value s for some real number s. Prove that there exists a point R such that

$$RA_1 + RA_2 + \cdots + RA_n < s.$$

Problem 1930.3.

Inside an acute triangle ABC is a point P that is not the circumcenter. Prove that among the segments AP, BP and CP, at least one is longer and at least one is shorter than the circumradius of ABC.

Problem 1943.2.

Let P be any point inside an acute triangle. Let D and d be respectively the maximum and minimum distances from P to any point on the perimeter of the triangle.

(a) Prove that $D \geq 2d$.

(b) Determine when equality holds.

Problem 1940.3.

(a) Prove that for any triangle H_1, there exists a triangle H_2 whose side lengths are equal to the lengths of the medians of H_1.

(b) If H_3 is the triangle whose side lengths are equal to the lengths of the medians of H_2, prove that H_1 and H_3 are similar.

Problem 1936.2.

S is a point inside triangle ABC such that the areas of the triangles ABS, BCS and CAS are all equal. Prove that S is the centroid of ABC.

Problem 1942.3.

Let A', B' and C' be points on the sides BC, CA and AB, respectively, of an equilateral triangle ABC. If $AC' = 2C'B$, $BA' = 2A'C$ and $CB' = 2B'A$, prove that the lines AA', BB' and CC' enclose a triangle whose area is $\frac{1}{7}$ that of ABC.

Problem 1929.3.

Let p, q and r be three concurrent lines in the plane such that the angle between any two of them is $60°$. Let a, b and c be real numbers such that $0 < a \leq b \leq c$.

(a) Prove that the set of points whose distances from p, q and r are respectively less than a, b and c consists of the interior of a hexagon if and only if $a + b > c$.

(b) Determine the length of the perimeter of this hexagon when $a + b > c$.

5.1   Discussion

5.1.1   Geometric Congruence and Inequalities

Geometry and number theory are the oldest branches of mathematics, treated in Euclid's monumental treatise, *The Elements*. Three of the 13 books of this work deal with number theory from a geometrical point of view, while the remaining 10 lay the foundation of synthetic geometry.

It is sad that geometry has lost much ground in the high school curriculum. People often overlook the fact that the subject nurtures visual perception and insight, although it is traditionally regarded as the subject to introduce deductive reasoning. It is a subject of surprising richness and yet accessible to students with little background other than some primitive notions about points, lines, the plane, distance, angles and polygons.

The classical instruments of Euclidean geometry are the **straightedge** and the **compass**. The former can be used to draw a straight line passing through two given points, while the latter can be used to draw a **circle**, which is the set of points at a fixed distance from a fixed point. The fixed distance is called the **radius** and the fixed point the **center** of the circle.

Angles are measured in degrees, denoted by °. The **complete angle** around one point is taken to have measure 360°. Half of a complete angle is called a **straight angle**, and half of a straight angle is called a **right angle**. The three points defining a straight angle are said to be **collinear** with one another while the two arms of a right angle are said to be **perpendicular** to each other.

Right away, we are ready to state and prove our first result.

Vertically Opposite Angles Theorem. *If the lines AB and CD intersect at a point O, then $\angle AOC = \angle BOD$.*

Proof. In Figure 5.1.1, observe that $\angle AOB$ and $\angle COD$ are straight angles. Hence

$$\angle AOC = \angle AOB - \angle COB$$
$$= 180° - \angle COB$$
$$= \angle COD - \angle COB$$
$$= \angle BOD.$$

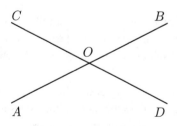

FIGURE 5.1.1

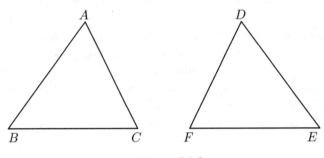

FIGURE 5.1.2

Geometry may be loosely defined as the study of shapes and sizes. Two triangles with the same shape and the same size are said to be **congruent** to each other. Geometric congruence is also an equivalence relation, just like the congruence relation in number theory.

If triangles ABC and DEF are congruent, as illustrated in Figure 5.1.2, then

(1) $BC = EF$;	(4) $\angle CAB = \angle FDE$;
(2) $CA = FD$;	(5) $\angle ABC = \angle DEF$;
(3) $AB = DE$;	(6) $\angle BCA = \angle EFD$.

Let us consider the converse problem of determining when two triangles are congruent. If it takes all six of the above statements to come to this conclusion, then congruence is not a very useful concept. Fortunately, there is a permanent half-price sale on selected items. If you buy three suitably chosen items from the six, you get the other three for free, and may make use of them later.

There are $\binom{6}{3} = 20$ possible subsets of size three, but they can be grouped into six types which we will denote by SSS, AAA, SAS, SSA, ASA and AAS. SSS means choosing (1), (2) and (3), the three statements about sides. The difference between SAS and SSA is that the chosen angle is between the two chosen sides in the former but not in the latter. For example, (1), (2) and (6) constitute an SAS case while (1), (2) and (4) constitute an SSA case.

It is commonly agreed that SAS should be taken as the basic case of congruence. One reason is that no matter what values are given to the three chosen items, a triangle always exists satisfying those conditions. Such is not the case for any of the other cases.

Basic principles in geometry are usually called *Postulates*, and so we have the **SAS Postulate**. Another one is the Triangle Inequality which we have already encountered in Chapter 2. It is a defining statement on how distance is to be measured in the plane or in higher dimensional spaces. We now start from the SAS Postulate and derive the other results in congruence, as well as some geometric inequalities.

ASA Theorem. *Triangles ABC and DEF are congruent if*

$$\angle CAB = \angle FDE, \quad AB = DE \quad \text{and} \quad \angle ABC = \angle DEF.$$

Proof. Note that if $BC = EF$, then the triangles are congruent by the SAS Postulate. Hence we may assume that $BC > EF$, so that there exists a point G on the segment BC such that $BG = EF$. Now triangles ABG and DEF are congruent by the SAS Postulate. It follows that we have $\angle CAB > \angle GAB = \angle FDE$, which contradicts $\angle CAB = \angle FDE$.

A triangle is said to be **isosceles** if at least two of its sides have the same length. If all three have the same length, it is then said to be **equilateral**.

Isosceles Triangle Theorem. *In triangle ABC, $\angle ABC = \angle BCA$ if and only if $CA = AB$.*

Proof. One part of the proof follows immediately from the SAS Postulate by comparing the triangles ABC and ACB. We have $AB = AC$. Since equality is symmetric, $AC = AB$. Finally, $\angle CAB = \angle BAC$ because they are the same angle. The other part can be proved in the same way using the ASA Theorem.

Sss Inequality. *If D is a point inside triangle ABC, then $BA + AC > BD + DC$.*

Proof. Extend BD to cut AC at E, as shown in Figure 5.1.3. By the Triangle Inequality, $DE + EC > DC$ and $BA + AE > BE = BD + DE$. Hence $AB + AC = AB + AE + EC > BD + CD$.

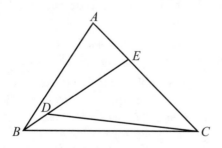

FIGURE 5.1.3

We now give the reason why we label this inequality Sss. As in the SAS Postulate, we are comparing two triangles. Here, they are ABC and DBC. More specifically, we are comparing three pairs of sides, BC with BC, CA with CD and AB with DB. This is why we use the letter s three times. Since $BC = BC$, we use the upper case. Since CA and CD are not necessarily equal, nor are AB and DB, we use lower cases. This is consistent with our earlier usage of the letters s and a in upper cases.

A result that complements the Isosceles Triangle Theorem essentially says that in a triangle, the longer side faces the bigger angle.

Angle-Side Inequality. *In triangle ABC, $\angle ABC > \angle ACB$ if and only if $AC > AB$.*

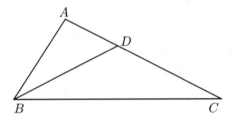

FIGURE 5.1.4

Proof. Suppose that $\angle ABC > \angle ACB$. Draw a line from B cutting AC at D, such that $\angle DBC = \angle ACB$, as shown in Figure 5.1.4. By the Isosceles Triangle Theorem, we have $BD = CD$. It follows from the Triangle Inequality that $AC = AD + CD = AD + BD > AB$. Conversely, suppose that $AC > AB$. Now exactly one of $\angle ABC < \angle ACB$, $\angle ACB = \angle ABC$ and $\angle ABC > \angle ACB$ must hold. The first case leads to $AC < AB$ by what we have just proved. The second case leads to $AC = AB$ by the Isosceles Triangle Theorem. It follows that the third case prevails.

SaS Inequality. *In triangles ABC and DEF, $AB = DE$ and $AC = DF$. Then $\angle BAC > \angle EDF$ if and only if $BC > EF$.*

Proof. Let G be the point on the same side of AB as C such that $AG = DF$ and $\angle BAG = \angle EDF$. Then triangles ABG and DEF are congruent by the SAS Postulate. Hence $EF = BG$. Suppose $\angle BAG > \angle BAC$. By the Isosceles Triangle Theorem, we have $\angle ACG = \angle AGC$. If B lies on the line CG, clearly we have $BG > BC$. Suppose B is on the same side of CG as A, as shown in the first diagram in Figure 5.1.5. Then

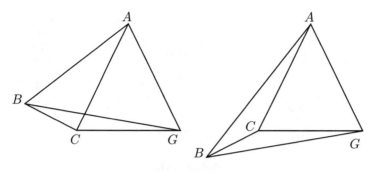

FIGURE 5.1.5

$\angle BCG > \angle ACG$ and $\angle AGC > \angle BGC$. By the Angle-Side Inequality, $BG > BC$. Suppose B is on the opposite side of CG as A, as shown in the second diagram in Figure 5.1.5. Then $\angle BCG > 180° - \angle ACG$ and $180° - \angle AGC > \angle BGC$. By the Angle-Side Inequality again, $BG > BC$. The converse can be proved using the same argument in the preceding result.

SSS Theorem. *Triangles ABC and DEF are congruent if $BC = EF$, $CA = FD$ and $AB = DE$.*

Proof.   Note that if $\angle ABC = \angle DEF$, then the triangles are congruent by the SAS Postulate. Hence we may assume that $\angle ABC > \angle DEF$. By the SaS Inequality, $CA > FD$, which contradicts $CA = FD$.

Let D be any point on the extension of the side BC of triangle ABC. Then $\angle ACD$ is called an **exterior angle** of the triangle. It may be greater than, equal to or less than the adjacent interior angle $\angle BCA$, but it is greater than either of the other two interior angles of the triangle.

Exterior Angle Inequality. *Let D be any point on the extension of the side BC of triangle ABC. Then $\angle ACD > \angle CAB$.*

Proof.   Since D is only used to determine $\angle ACD$, we may choose D so that $CD = AB$, as illustrated in Figure 5.1.6. By the Triangle Inequality, $DA + AB > BD = BC + CD$, which simplifies to $DA > BC$. Note that in triangles CAB and ACD, we have $CA = AC$, $AB = CD$ while $BC < DA$. By the SaS Inequality, $\angle ACD > \angle CAB$.

Let P be a point not on a line AB. How do we measure the distance from P to AB? We want it to be the distance between P and the point Q on AB nearest to P. We prove that this point Q is such that PQ is perpendicular to AB. For any other point R on AB, we have $\angle PQR = 90°$, so that it is

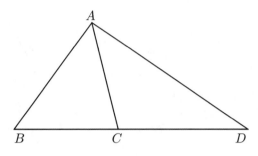

FIGURE 5.1.6

equal to the adjacent exterior angle of triangle PQR. This angle is greater than $\angle PRQ$ by the Exterior Angle Inequality. Hence $\angle PQR > \angle PRQ$, and we have $PR > PQ$ by the Angle-Side Inequality.

AAS Theorem.   *Triangles ABC and DEF are congruent if*

$$\angle CAB = \angle FDE, \quad \angle ABC = \angle DEF \quad \text{and} \quad BC = EF.$$

Proof.   Note that if $AB = DE$, then the triangles are congruent by the SAS Postulate. Hence we may assume that $AB > DE$, so that there exists a point G on AB such that $GB = DE$. Now triangles GBC and DEF are congruent by the SAS Postulate. It follows that $\angle CGB = \angle FDE = \angle CAB$. However, this contradicts the Exterior Angle Inequality since $\angle CGB$ is an exterior angle of triangle CAG and $\angle CAB$ is one of the interior angles not adjacent to $\angle CGB$.

As an application of the AAS Theorem, we give a second solution to **Problem 1935.2**. Recall that we have already solved the one-dimensional version of it in subsection 2.1.4, and our approach now is to reduce the problem to one dimension.

Problem 1935.2.

Prove that a finite point set cannot have more than one center of symmetry.

Second Solution: Suppose O_1 and O_2 are distinct centers of symmetry of a finite point set S. For any point P in S, let P' be the point on the line $O_1 O_2$ such that PP' is perpendicular to $O_1 O_2$. Let S' be the set consisting of such projections of the points of S onto the line $O_1 O_2$. We claim that O_1 and O_2 are still centers of symmetry of S'. This will then contradict the one-dimensional case of the problem. Let P and Q be points in S symmetric about O_1. Consider their projections P' and Q' in

S'. In triangles O_1PP' and O_1QQ', we have $O_1P = O_1Q$. We also have $\angle PO_1P' = \angle QO_1Q'$ by the Vertically Opposite Angles Theorem. Finally, $\angle O_1P'P = 90° = \angle O_1Q'Q$. Hence these two triangles are congruent by the AAS Theorem, so that $O_1P' = O_2Q'$. Thus P' and Q' are symmetric about O_1. It follows that O_1 is indeed a center of symmetry of S'. Similarly, so is O_2, justifying our claim.

Can we prove triangle congruence in the remaining two cases, namely, AAA and SSA? If the three angles of one triangle are equal to the corresponding ones in another triangle, then the triangles must have the same shape, but not necessarily the same size. Hence they need not be congruent to each other. However, all is not lost. We can salvage a partial result later in subsection 5.1.4.

The SSA case is the most interesting. Let us first point out that it does not always lead to congruence. In Figure 5.1.7, triangles ABC and ABD are not congruent. However, $AB = AB$, $\angle ABC = \angle ABD$ and $AC = AD$.

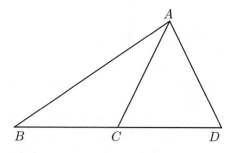

FIGURE 5.1.7

Suppose we are given AB, $\angle ABC$ and AC, and are asked to construct triangle ABC. We can start with the side AB and the line ℓ on which the side BC lies. We can draw this line since we know $\angle ABC$, even though we do not know BC. To construct the point C, draw a circle with center A and radius AC. Let us first suppose that $\angle ABC < 90°$.

If AC is too short, the circle will not cut ℓ, and triangle ABC does not exist. If it is just long enough to touch ℓ, we will have a unique solution to the construction problem, and $\angle BCA = 90°$. If AC is longer than that but shorter than AB, we will have two solutions as shown in the diagram above. Finally, if it is longer than or equal to AB, we will have a unique solution once again.

If $\angle ABC = 90°$, then we must have $AC > AB$. There will be two solutions, but they are congruent to each other and need not be distinguished. If $\angle ABC > 90°$, we must have $AC > AB$ again, and the solution will be unique. So, when we encounter the SSA case, we should not jump to any conclusion but analyse the situation carefully. Nevertheless, there is a special case that we can isolate.

HSR Theorem. *Triangles ABC and DEF are congruent if $\angle BCA = 90° = \angle EFD$, $AB = DE$ and $AC = DF$.*

Proof. Extend BC to G so that $CG = FE$. Then triangles AGC and DEF are congruent by the SAS Theorem. Hence $AG = DE = AB$ and $\angle ABC = \angle AGC$ by the Isosceles Triangle. It follows that triangles ABC and AGC are congruent by the AAS Theorem, and the desired conclusion follows.

The letter H stands for **hypotenuse**, a name given to the side facing the right angle.

5.1.2   Parallelism

Two lines are said to be **parallel** to each other if they do not intersect, no matter how far they are extended. We use the notation $\ell_1 \parallel \ell_2$ to denote the fact that the lines ℓ_1 and ℓ_2 are parallel. We would like parallelism to be an equivalence relation. That it is reflexive is by convention, and the definition makes it clear that it is symmetric. However, that it is transitive is not at all obvious. In fact, we have to introduce it as the **Parallel Postulate**.

Let P be a point not on a line ℓ. Through P, we can draw a line m perpendicular to ℓ, cutting it at L. Through P, we can draw another line n perpendicular to m. We claim that n is parallel to ℓ. Otherwise, let them intersect at some point A. Let Q be any point on the extension of AP. Then both $\angle ALP$ and $\angle LPQ$ are right angles and therefore equal to each other. However, by the Exterior Angle Inequality, $\angle LPQ > \angle ALP$, which is a contradiction. It follows that through P, we can draw at least one line parallel to ℓ.

Playfair's Theorem. *Through a point P not on a line ℓ, at most one line parallel to ℓ can be drawn.*

Proof. Let ℓ_1 and ℓ_2 be lines through P parallel to ℓ. Since ℓ_2 is parallel to ℓ and ℓ is parallel to ℓ_1, ℓ_2 is parallel to ℓ_1 by the Parallel Postulate. Since they intersect at the point P, they must be the same line.

Figure 5.1.8 shows a line cutting two other lines AB and CD at G and H respectively. Let E be any point on the extension of HG and F be any

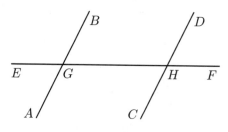

FIGURE 5.1.8

point on the extension of GH. Then the angles $\angle AGE$ and $\angle CHG$ are called **corresponding angles**.

Corresponding Angles Theorem. A line cuts two other lines to form a pair of corresponding angles. These angles are equal if and only if those two lines are parallel to each other.

Proof. We will use the notation in Figure 5.1.8. Suppose $\angle AGE = \angle CHG$. If the extension of BA and DC meet at some point K, then $\angle AGE$ is an exterior angle and $\angle CHG$ is a non-adjacent interior angle of triangle GHK. By the Exterior Angle Theorem, $\angle AGE > \angle CHG$. On the other hand, if the extension of AB and CD meet at some point L, then $\angle DHE$ is an exterior angle and $\angle BGH$ is a non-adjacent interior angle of triangle GHL. By the Vertically Opposite Angles Theorem and the Exterior Angle Inequality, $\angle AGE = \angle BGH < \angle DHE = \angle CHG$. It follows that AB and CD must be parallel to each other. Conversely, if AB and CD are parallel, then those two angles must be equal. Otherwise, we can draw a line $C'D'$ through H such that $\angle C'HG = \angle AGE$. Then $C'D'$ will also be parallel to AB, contradicting Playfair's Theorem.

The angles $\angle BGH$ and $\angle CHG$ in Figure 5.1.8 are called **alternate angles** of AB and CD with respect to EF.

Alternate Angles Theorem. *A line cuts two other lines to form a pair of alternate angles. These angles are equal if and only if those two lines are parallel to each other.*

Proof. This follows easily from the Vertically Opposite Angles Theorem and the Corresponding Angles Theorem.

When an example of a result in geometry is requested, one of the most popular answers is the following.

Angle Sum Theorem. *The sum of the three angles of a triangle is $180°$.*

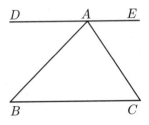

FIGURE 5.1.9

Proof. Draw a line DE through vertex A of an arbitrary triangle ABC parallel to BC, as shown in Figure 5.1.9. By the Alternate Angles Theorem, $\angle ABC = \angle BAD$ and $\angle BCA = \angle EAC$. Note that we have $\angle BAD + \angle CAB + \angle EAC = 180°$. Hence the sum of the three angles of triangle ABC is also $180°$.

Exterior Angle Theorem. *An exterior angle of a triangle is equal to the sum of the two non-adjacent interior angles of the triangle.*

Proof: This follows easily from the Angle Sum Theorem.

Obviously, no two sides of a triangle can be parallel to each other. To have two parallel sides, a polygon must have at least four sides. In such a polygon, it is possible for a line to join two vertices without being a side. Such a line is called a **diagonal** of the polygon. It is also possible for two sides to intersect, but we assume that this does not happen. By convention, we label the vertices of a polygon cyclically.

A quadrilateral is a polygon with four sides. If it has two pairs of parallel sides, then it is called a **parallelogram**. Let $ABCD$ be a parallelogram whose diagonals intersect at the point E. Then

(1) $AB \parallel CD$;   (5) $\angle ABC = \angle CDA$;
(2) $AD \parallel BC$;   (6) $\angle BCD = \angle DAB$;
(3) $AB = CD$;   (7) $AE = CE$;
(4) $AD = BC$;   (8) $BE = DE$.

Both (1) and (2) come straight from the definition. The others can be proved easily from various pairs of congruent triangles in the parallelogram. As for the converse problem of how many of these items are sufficient to guarantee that $ABCD$ is a parallelogram, we have an even bigger discount than in the case of congruent triangles. Two suitably chosen items will do, for a saving of seventy-five percent.

Obviously, (1) and (2) will do. Other workable combinations are (3) and (4), (5) and (6), as well as (7) and (8). We can also mix and match, such as (1) and (3). We give a proof of this last case. Since AB is parallel to CD, $\angle ABD = \angle CDB$ by the Alternate Angles Theorem. Obviously, $BD = DB$. Along with $AB = CD$, they imply that triangles ABD and CDB are congruent. We have $AD = BC$ and $\angle ADB = \angle CBD$. The latter implies that AD and BC are parallel by the Alternate Angles Theorem. The other four working pairs listed above can be proved in an analogous way.

One must not come away with the impression that any two of these eight items imply that $ABCD$ is a parallelogram. In fact, most combinations do not work. We omit the detailed analysis.

A set of points defined by a given condition is called a **locus**. For example, the locus of a point at a fixed distance r from the a fixed point O is the circle with center O and radius r. What is the locus of a point at a fixed distance from a line? Let ℓ be a line and P be a point not on it. First, we claim that any point on the line through P parallel to ℓ lies on the locus. Let Q be such a point. Draw lines through P and Q perpendicular to ℓ, cutting it at S and R respectively. Then $PQRS$ is a parallelogram, so that $PS = QR$.

It is easy to prove that any point between PQ and ℓ has smaller distance to ℓ than P, and any point on the opposite side of PQ to ℓ has greater distance. However, we must not conclude hastily that the locus is the line PQ. By symmetry, there is another line parallel to ℓ on the other side that is also part of the locus.

Midpoint Theorem. *The segment joining the midpoints of two sides of a triangle is parallel to the third side and equal to half its length.*

Proof. Let E and F be the respective midpoints of the sides CA and AB of an arbitrary triangle ABC. Extend FE to D so that $FE = ED$, as shown in Figure 5.1.10. Now $\angle AEF = \angle CED$ by the Vertically Opposite

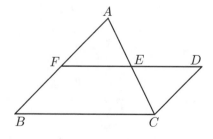

FIGURE 5.1.10

Angles Theorem. Along with $EA = EC$, they imply that triangles AEF and CED are congruent. Hence $CD = AF = BF$ and $\angle ECD = \angle EAF$. By the Alternate Angles Theorem, CD is parallel to BF. Hence $BCDF$ is a parallelogram, so that $BC = FD$ and BC is parallel to FE. Since $FE = ED$, we indeed have $FE = \frac{1}{2}BC$.

As an application of this result, we give a third solution to **Problem 1935.2**.

Problem 1935.2.

Prove that a finite point set cannot have more than one center of symmetry.

Third Solution: Using the Midpoint Theorem, we can prove the stronger result that if a point set S has two distinct centers O_1 and O_2 of symmetry, then it is an infinite set and has infinitely many centers of symmetry. Let P_0 be a point in S. We first assume that P_0 is not on the line O_1O_2 as illustrated in Figure 5.1.11. For $k \geq 0$, let P_{2k+1} be obtained by reflecting P_{2k} across O_1, and let P_{2k+2} be obtained by reflecting P_{2k+1} across O_2. Then $P_{2k}P_{2k+2}$ is parallel to O_1O_2 and equal to twice its length. It follows that P_4 is on the extension of P_0P_2, P_6 is in the extension of P_0P_4, and so on. Since all these points belong to S, S is not finite. It is easy to see that all points on the line O_1O_2 such that their distances from O_1 are integral

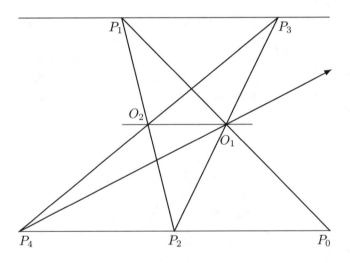

FIGURE 5.1.11

multiples of O_1O_2 are also centers of symmetry of S. The case where all points in S are on the line O_1O_2 can be handled in a similar manner.

5.1.3   Centers of a Triangle

A line joining a vertex of a triangle to a point on the opposite side or its extension is called a **cevian** of the triangle. A cevian is called a **median** if the point on the opposite side is its midpoint, and an **altitude** if it is perpendicular to the opposite side. It may also be an **angle bisector**, dividing the angle at the given vertex into two equal parts. We first prove an important property about medians.

Median Trisection Theorem.   *Let AD and BE be medians of triangle ABC such that they intersect each other at the point G. Then $AG = 2GD$ and $BG = 2GE$.*

Proof.   Suppose first that AD and BE are medians. Let P and Q be the respective midpoints of AG and BG, as illustrated in Figure 5.1.12. By the Midpoint Theorem, both PQ and ED are parallel to AB and equal to half its length. Hence they are parallel and equal to each other, so that $DEPQ$ is a parallelogram. Now $DG = GP = PA$ and $AG = 2DG$. Similarly, $BG = 2EG$.

   Lines passing through the same point are said to be **concurrent**. Let AD, BE and CF be cevians of triangle ABC. Consider the following four statements:

1. AD is a median;
2. BE is a median;
3. CF is a median;
4. AD, BE and CF are concurrent.

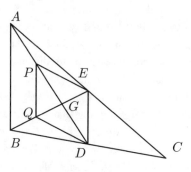

FIGURE 5.1.12

It turns out that if any three of them is true, then so is the fourth. By symmetry, we may assume that (1) and (2) are true. Then AD and BE intersect at a point G with $AG = 2DG$ by the Median Trisection Theorem. Suppose (3) is also true. Then AD and CF must also intersect at G because $AG = 2DG$. Hence (4) is also true. Conversely, if CF passes through G but is not a median, then the median CF' must also pass through G. This is a contradiction. The point where the three medians meet is called the **centroid** of the triangle.

In the above set of statements, if we replace the term "median" by the term "angle bisector," the four-for-three sale is still on. We first give a characterization of the bisector of an angle.

Two intersecting lines form four angles. We claim that the locus of the points equidistant from the two lines consists of two mutually perpendicular lines which are the bisectors of the angles formed by the lines. To establish this fact, we have to prove two things. First, let P be a point equidistant to two lines ℓ_1 and ℓ_2 intersecting at a point O. Draw lines through P perpendicular to ℓ_1 and ℓ_2, cutting them at Q and R respectively. Then we have $PQ = PR$, $PO = PO$ and $\angle PQO = 90° = \angle PRO$. Hence triangles PQO and PRO are congruent by the HSR Theorem, so that $\angle POQ = \angle POR$. It follows that P lies on the bisector of $\angle QOR$. Conversely, let P be on the bisector of one of the angles formed by ℓ_1 and ℓ_2. Construct Q and R as before. Now triangles PQO and PRO are still congruent to each other, but by the AAS Theorem. Then $PQ = PR$, and P indeed lies on the locus.

Returning to the four statements with "median" replaced by "angle bisector," we may assume as before that (1) and (2) are true. Let AD and BE intersect at a point I. Since it is on the bisector of $\angle CAB$, it is equidistant from CA and AB. Since it is on the bisector of $\angle ABC$, it is equidistant from AB and BC. It follows that I is equidistant from BC and CA. It is easy to see that (3) and (4) will imply each other.

The point I where the three angle bisectors meet is called the **incenter** of the triangle. The common value r of its distance to the sides is called the **inradius** of the triangle. This is because the circle with center I and radius r will touch the three sides of the triangle. It is called the **incircle** of the triangle, and is the largest circle that will fit inside the triangle.

Let H be a point on the extension of AB and K be a point on the extension of AC. Then the bisector of the interior angle $\angle CAB$ and the exterior angles $\angle CBH$ and $\angle BCK$ are also concurrent. The point of concurrency is called an **excenter** of the triangle, and its distance to BC

or the extension of either AB or AC is called the corresponding **exradius** of the triangle.

We can draw a circle touching the side BC and the extensions of the sides AB and AC. It is called an **excircle** of the triangle. There are two other excircles. The center of one of them is the point of concurrency of the bisectors of the interior angle at B and the exterior angles at C and A. The center of the other one is the point of concurrency of the bisectors of the interior angle at C and the exterior angles at A and B.

The concept of equidistance gives rise to another interesting locus. Let A and B be fixed points. We wish to find the locus of a point P equidistant from A and B. Clearly, the midpoint M of AB lies on the locus. If P is another point on the locus, then we have $PA = PB, MA = MB$ and $PM = PM$. Hence triangles PAM and PMB are congruent by the SSS Theorem, so that $\angle PMA = \angle PMB$. Since they sum to $180°$, each is a right angle. Hence P lies on the line through M perpendicular to AB. It is called the **perpendicular bisector** of AB.

We still have to prove that any point Q not on the perpendicular bisector of AB does not belong to the locus. We may assume that Q is on the same side of this line as A, as shown in Figure 5.1.13. Let BQ cut the perpendicular bisector at a point P. Then $PA = PB$. By the Triangle Inequality, $BQ = BP + PQ = AP + PQ > AQ$, so that Q is indeed not on the locus.

Let the perpendicular bisectors of the sides AB and BC of an arbitrary triangle ABC meet at a point O. Then O is equidistant from A and B as well as equidistant from B and C. Hence it is equidistant from C and A, and lies on the perpendicular bisector of CA. Thus we have yet another four-for-three sale, though not involving cevians this time.

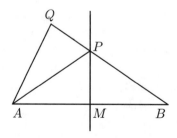

FIGURE 5.1.13

The point O of concurrency of the perpendicular bisectors of the sides of triangle ABC is called the **circumcenter** of ABC, and the common value R of its distance to the vertices is called the **circumradius**. The circle with center O and radius R is called the **circumcircle** of the triangle, and it passes through all three vertices. Note that the circumcircle is not necessarily the smallest circle that contains the triangle. For example, if BC is almost equal to the combined lengths of AB and AC, the circumcircle of triangle ABC is much larger than the circle with BC as diameter. Also, unlike the centroid or the incenter, the circumcenter may lie outside the triangle or on its perimeter.

We now consider the altitudes of a triangle. In the four statements about the medians, we may also replace the term "median" with the term "altitude" and have yet another four-for-three sale.

Through the vertices of triangle ABC, draw lines parallel to the opposite sides, forming a larger triangle XYZ. Now A, B and C are the midpoints of the sides of XYZ. Hence the altitudes of ABC are just the perpendicular bisectors of the sides of XYZ, which are already known to be concurrent. This point of concurrency is called the **orthocenter** of triangle ABC, and it does not necessarily lie inside the triangle.

We now offer a close-out sale for this subsection in a block-buster seven-for-one deal. We claim that if one of the following statements is true, then so are the others.

    (1)   The triangle is equilateral.

    (2)   The centroid and the incenter of the triangle coincide.

    (3)   The centroid and the circumcenter of the triangle coincide.

    (4)   The centroid and the orthocenter of the triangle coincide.

    (5)   The incenter and the circumcenter of the triangle coincide.

    (6)   The incenter and the orthocenter of the triangle coincide.

    (7)   The circumcenter and the orthocenter of the triangle coincide.

Clearly, if (1) is true, so are all the others. Hence it is sufficient to prove that (1) follows from any of the other six statements. Moreover, it is only necessary to prove that the triangle is isosceles, since by symmetry it will have to be equilateral.

To prove that (2) implies (1), let AD be a median of triangle ABC. It is also the bisector of $\angle CAB$. Extend AD to Y so that $AD = DY$. Then $ABYC$ is a parallelogram since its diagonals bisect each other. Hence $\angle BYD = \angle CAD = \angle BAD$ by the Alternate Angles Theorem, so that $AB = BY = AC$ as desired.

Next, we prove that (3) implies (1). Let AD be a median of triangle ABC. Since it passes through the circumcenter and the midpoint of BC, it is perpendicular to BC. We have $BD = CD$, $AD = AD$ and $\angle BDA = 90° = \angle CDA$. Hence triangles BAD and CAD are congruent by the SAS Postulate, so that $AB = AC$ as desired.

To prove that (4) implies (1), let AD be an altitude of triangle ABC. It is also a median. We have $BD = CD$, $AD = AD$ and $\angle BDA = 90° = \angle CDA$. Hence triangles BAD and CAD are congruent by the SAS Postulate, so that $AB = AC$ as desired.

We now prove that (5) implies (1). Let X be the incenter as well as the circumcenter of triangle ABC. Let E and F be the respective midpoints of CA and AB. Then $\angle FAX = \angle EAX, AX = AX$ and $\angle XFA = 90° = \angle XEA$. Hence triangles FAX and EAX are congruent by the HSR Theorem. It follows that $AB = 2AF = 2AE = AC$.

To prove that (6) implies (1), let AD be an altitude of triangle ABC. It is also the bisector of $\angle CAB$. We have $\angle BAD = \angle CAD$, $AD = AD$ and $\angle BDA = 90° = \angle CDA$. Hence triangles BAD and CAD are congruent by the ASA Theorem, so that $AB = AC$ as desired.

Finally, we prove that (7) implies (1). Let AD be an altitude of triangle ABC. Since it passes through the circumcenter, D is the midpoint of BC. Now $BD = CD$, $\angle BDA = 90° = \angle CDA$ and $AD = AD$. Hence triangles BAD and CAD are congruent by the SAS Postulate, so that $AB = AC$ as desired.

5.1.4 Area and Similarity

The word *geometry* means measurement (*metron*) of the earth (*geo*). In ancient time, people's wealth was often expressed in terms of the amount of land they owned, and geometry was a practical subject. Nowadays, the measurement of area is still probably the earliest introduction a child has to geometry, often encountered in the later stage of elementary school. Every polygon P is assumed to have a positive area which is denoted by $[P]$. A basic notion is that congruent figures have the same area.

A **rectangle** is a parallelogram with a right angle. In fact, all four angles will be right angles. This is taken to be our basic figure in the study of area. Its area is given by the product of the lengths of two adjacent sides, often called its *length* and *width*. A special case is a **square**, which is a rectangle with two adjacent sides equal. In fact, all four sides will be equal. Its area is equal to the square of a side.

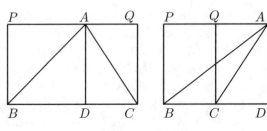

FIGURE 5.1.14

From the basic formula for the area of a rectangle, we can deduce that the area of a triangle is given by half the product of one side and the altitude to that side. This side is called the **base**.

Let AD be the altitude to the chosen base BC. Suppose D is between B and C, as illustrated in the first diagram in Figure 5.1.14. Construct the rectangle $BCQP$ such that A is between P and Q. Note that $[ABC] = [BAD] + [CAD]$, $[ABP] = [BAD] = \frac{1}{2}[ADBP]$ and $[ACQ] = [CAD] = \frac{1}{2}[ADCQ]$. Hence

$$[ABC] = \frac{1}{2}[BCPQ] = \frac{1}{2}BC \cdot CQ = \frac{1}{2}BC \cdot AD.$$

Suppose D is on the extension of BC, as illustrated in the second diagram in Figure 5.1.14. The statement above still holds if we replace the plus signs with minus signs.

Angle Bisector Theorem. *If the cevian AD of triangle ABC is the bisector of $\angle CAB$, then $\frac{BD}{CD} = \frac{AB}{AC}$.*

Proof.   Both ratios in the conclusion are equal to $\frac{[BAD]}{[CAD]}$. If we regard BD and CD as the respective bases of the triangles, then they have the same altitude which is the distance from A to BC. If we regard AB and AC as the respective bases instead, the respective altitudes are the distances from D to AB and AC. Since D lies on the bisector of $\angle CAB$, it is equidistant from its two arms.

We can give an alternative proof that if the centroid and the incenter of a triangle coincide, then it must be isosceles, and in fact equilateral. Let AD be a median of triangle ABC. Then it is also the bisector of $\angle CAB$. From the Angle Bisector Theorem, $\frac{AB}{AC} = \frac{BD}{CD} = 1$ since D is the midpoint of BC. It follows that $AB = AC$.

Equilateral Triangle Altitude Theorem.   *The total distance from any point inside an equilateral triangle to the three sides is equal to the altitude of the triangle.*

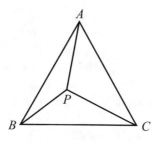

FIGURE 5.1.15

Proof. Let P be any point inside an equilateral triangle ABC. We may take the side length to be s. Let the respective distances from P to BC, CA and AB be x, y and z. Let h denote the altitude of ABC. Then $x + y + z = h$ since

$$(x + y + z)s = 2([PCB] + [PAC] + [PBA])$$
$$= 2[ABC] = hs.$$

Pythagoras' Theorem. *In triangle ABC, if $\angle BCA = 90°$, then $AB^2 = BC^2 + CA^2$.*

Proof. Let the squares $ABDE$, $BCFG$ and $CAHK$ be constructed outside triangle ABC, as shown in Figure 5.1.16. Since $\angle BCA = 90°$,

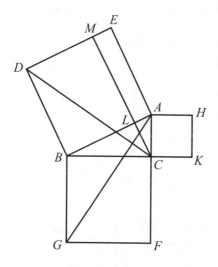

FIGURE 5.1.16

A, C and F are collinear, as are B, C and K. Draw a line through C perpendicular to AB, cutting it at L and DE at M. In triangles BCD and BGA, we have $BD = BA$ and $BC = BG$. Moreover, we have $\angle DBC = 90° + \angle ABC = \angle ABG$. Hence the two triangles are congruent by the SAS Postulate, and have the same area. Considering BD as the base for triangle BCD, its altitude is the distance from C to BD, which is equal to MD. Hence $[BCD] = \frac{1}{2}BD \cdot MD = \frac{1}{2}[BDML]$. Similarly, $[BGA] = \frac{1}{2}[BCFG]$. Hence $[BDML] = [BCFG]$. In an analogous way, we can prove that $[AEML] = [CAHK]$. It follows that $[ABDE] = [BDML] + [AEML] = [BCFG] + [CAHK]$, so that $AB^2 = BC^2 + CA^2$.

This is certainly the most quoted result in geometry. The proof above is attributed to Euclid himself. The converse is often expressed in the form of inequalities.

Pythagoras' Inequality.   *If $\angle BCA < 90°$, then $AB^2 < BC^2 + CA^2$. If $\angle BCA > 90°$, then $AB^2 > BC^2 + CA^2$.*

Proof.   Let DEF be a triangle with $EF = BC$, $FD = CA$ and $\angle EFD = 90°$. By Pythagoras' Theorem, we have $DE^2 = EF^2 + FD^2$. Suppose $\angle BCA < 90° = \angle EFD$. By the SaS Inequality, $AB < DE$, so that

$$AB^2 < DE^2 = EF^2 + FD^2 = BC^2 + CA^2.$$

The proof for the case where $\angle BCA > 90°$ is analogous.

Median Theorem.   *Let AD be a median of triangle ABC. Then*

$$AB^2 + AC^2 = 2AD^2 + 2BD^2.$$

Proof.   Drop the perpendicular AK from A to BC, as shown in Figure 5.1.17. By Pythagoras' Theorem,

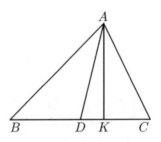

FIGURE 5.1.17

$$AB^2 + AC^2 = BK^2 + 2AK^2 + CK^2$$
$$= (BD + DK)^2 + 2AK^2 + (CD - DK)^2$$
$$= 2BD^2 + 2(AK^2 + DK^2)$$
$$= 2BD^2 + 2AD^2.$$

The argument is similar if K is not between B and C.

Recall from Section 5.1 that the AAA case does not lead to congruence. If two triangles have the same angles, then they have the same shape but not necessarily the same size. They are said to be **similar** to each other. Like congruence, similarity is an equivalence relation.

If triangles ABC and DEF are similar, then

(1) $\angle CAB = \angle FDE$;

(2) $\angle ABC = \angle DEF$;

(3) $\angle BCA = \angle EFD$;

(4) $\dfrac{CA}{AB} = \dfrac{FD}{DE}$;

(5) $\dfrac{AB}{BC} = \dfrac{DE}{EF}$;

(6) $\dfrac{BC}{CA} = \dfrac{EF}{FD}$.

As in the case of congruent triangles, it is not necessary to prove all six items in order to conclude that ABC and DEF are similar triangles. It turns out that a suitable choice of two of them will work.

AA Theorem.   *If two angles of one triangle are equal respectively to two angles of another triangle, then these two triangles are similar to each other.*

Proof.   By the Angle Sum Theorem, The third angles of one triangle must be equal to the third angle of the other triangle. Hence the two triangles are similar to each other.

sAs Theorem.   *If an angle of one triangle is equal to an angle of another triangle, and the sides enclosing those angles are in the same proportion in both triangles, then the two triangles are similar to each other.*

Proof.   Suppose that in triangles ABC and DEF, $\angle CAB = \angle FDE$ and $\frac{AB}{DE} = \frac{CA}{FD} = k$ for some positive number k. If $k = 1$, the two triangles are in fact congruent to each other. By symmetry, we may assume that $k > 1$. Take H on AB so that $AH = DE$, and take G on AC so that $AG = DF$, as shown in Figure 5.1.18. Now

$$\frac{[ABC]}{[AHC]} = \frac{AB}{AH} = \frac{AB}{DE} = k.$$

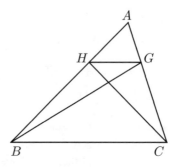

FIGURE 5.1.18

Similarly,

$$\frac{[ABC]}{[ABG]} = \frac{AC}{AG} = \frac{AC}{DF} = k.$$

Hence $[AHC] = \frac{1}{k}[ABC] = [ABG]$, so that

$$[BCH] = [ABC] - [AHC] = [ABC] - [ABG] = [BCG].$$

Now triangles BCG and BCH have a common base BC. Since they have equal area, they must have equal altitude to this base. It follows that HG is parallel to BC. By the Corresponding Angles Theorem, $\angle AHG = \angle ABC$. By the AA Theorem, triangles ABC and AHG are similar. Since AHG and DEF are congruent by the SAS Theorem, ABC and DEF are similar to each other.

sss Theorem. *If the three sides of one triangle are in the same proportion as the three sides of another triangle, then the two triangles are similar.*

Proof. Let k denote the common value of $\frac{AB}{DE} = \frac{BC}{EF} = \frac{CA}{FD}$. Suppose $k = 1$. Then triangles ABC and DEF are congruent by the SSS Theorem. Hence we may assume that $k > 1$. As in Figure 5.1.18, take H on AB and G on AC such that $AH = DE$ and $AG = DF$. By the sAs Theorem, ABC and AHG are similar. Hence

$$\frac{DE}{EF} = \frac{AB}{BC} = \frac{AH}{HG} = \frac{DE}{HG}.$$

It follows that $HG = EF$, so that AHG and DEF are congruent by the SSS Theorem. Hence ABC and DEF are similar.

We now give some simple applications of similar triangles. In triangle ABC, let E be a point on CA and F a point on AB such that FE is parallel to BC and equal to half its length. By the Corresponding Angles

Theorem, $\angle AFE = \angle ABC$ and $\angle AEF = \angle ACB$. Hence triangles AFE and ABC are similar by the AA Theorem. From $\frac{EA}{CA} = \frac{AF}{AB} = \frac{FE}{BC} = \frac{1}{2}$, we conclude that E is the midpoint of CA and F is the midpoint of AB. This is the converse of the Median Theorem.

Let AD and BE be cevians of triangle ABC such that they intersect each other at the point G. Moreover, $AG = 2GD$ and $BG = 2GE$. Let P be the midpoint of AG and Q be the midpoint of BG, as illustrated in Figure 5.1.12. Then $PG = GD$ and $QG = GE$, so that $DEPQ$ is a parallelogram. Hence DE is parallel to QP and they are equal in length. By the Median Theorem, QP is parallel to AB and equal to half its length. It follows that so is DE. By the converse of the Median Theorem, D is the midpoint of BC and E is the midpoint of CA. This is the converse of the Median Trisection Theorem.

Euler's Inequality. *The circumradius of any triangle is greater than or equal to twice its inradius.*

Proof. Let ABC be any triangle and let D, E and F be the respective midpoints of BC, CA and AB. By the Midpoint Theorem, each side of triangle DEF is equal to half the length of the corresponding sides of triangle ABC. Hence the two triangles are similar by the sss Theorem. Moreover, the circumcircle of triangle DEF is exactly half the size of the circumcircle of triangle ABC. Now the smaller circle is one that intersects all three sides of triangle ABC. The incircle of ABC is the smallest such circle since it fits inside ABC. It follows that the circumradius of a triangle is greater than or equal to twice its inradius. Equality holds if and only if the triangle is equilateral.

Intercept Theorem. *Suppose a line intersects three parallel lines at the points B, C and D, and another line intersects the same three parallel lines at the points E, F and G, respectively. Then $\frac{BC}{CD} = \frac{EF}{FG}$.*

Proof. If BD is parallel to EG, then $BCFE$ and $CDGF$ are parallelograms. We have in this case $BC = EF$ and $CD = FG$. Suppose that BD and EG intersect at a point A, as shown in Figure 5.1.19. Then ABE, ACF and ADG are all similar to one another by the AA Theorem. Let k denote the common value of $\frac{AB}{AE} = \frac{AC}{AF} = \frac{AD}{AG}$. Note that we have $BC = AC - AB = k(AF - AE) = kEF$ and $CD = AD - AC = k(AG - AF) = kFG$. Hence $\frac{BC}{EF} = k = \frac{CD}{FG}$, and the desired result follows immediately.

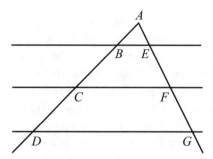

FIGURE 5.1.19

5.2 Solutions

Problem 1931.3.

Let A and B be two given points, distance 1 apart. Determine a point P on the line AB such that $\frac{1}{1+AP} + \frac{1}{1+BP}$ is a maximum.

First Solution: Let $f(P) = \frac{1}{1+PA} + \frac{1}{1+PB}$. If P is not on the segment AB, we may assume that A lies between P and B. Then $PA > 0$ and $PB > AB$ so that $f(P) < 1 + \frac{1}{1+AB} = f(A)$. It follows that the maximum value of $f(P)$ occurs when P is on the segment AB. Let C be on the extension of BA and D be on the extension of AB such that $CA = 1 = BD$. Let M be the midpoint of CD. Then

$$
\begin{aligned}
f(P) &= \frac{1}{CP} + \frac{1}{DP} \\
&= \frac{CD}{CP \cdot DP} \\
&= \frac{CD}{(CM \pm PM)(DM \mp PM)} \\
&= \frac{CD}{CM^2 - PM^2}.
\end{aligned}
$$

Now CD and CM are constants. Hence $f(P)$ is a maximum if and only if PM is a maximum. This occurs precisely when $P = A$ or $P = B$.

Second Solution: As in the First Solution, we may restrict our attention to the segment AB. Let $PA = x$, where $0 \le x \le 1$. Then

$$
\frac{1}{1+PA} + \frac{1}{1+PB} = \frac{1}{1+x} + \frac{1}{2-x} = \frac{3}{2+x-x^2}.
$$

Hence the maximum value occurs when $2 + x(1 - x)$ is a minimum. Clearly, this occurs at $x = 0$ or $x = 1$ where $x(1 - x) = 0$. This is because $x(1 - x) > 0$ when $0 < x < 1$. Hence the maximum value $\frac{3}{2}$ of the given expression occurs when $P = A$ or $P = B$.

Problem 1942.1.

Prove that in any triangle, at most one side can be shorter than the altitude from the opposite vertex.

Solution: Suppose AD and BE are the altitudes of triangle ABC which are longer than BC and CA respectively. In triangle ADC, we have $\angle ADC = 90° > \angle ACD$. By the Angle-Side Inequality, $CA > AD$. Similarly, in triangle BCE, $BC > BE$. It follows that we have

$$AD > BC > BE > CA > AD,$$

which is a contradiction. Hence in any triangle, at most one side can be shorter than the altitude from the opposite vertex.

Problem 1937.3.

Let n be a positive integer. Let P, Q, A_1, $A_2, \ldots,$ A_n be distinct points such that A_1, $A_2, \ldots,$ A_n are not collinear. Suppose that $PA_1 + PA_2 + \cdots + PA_n$ and $QA_1 + QA_2 + \cdots + QA_n$ have a common value s for some real number s. Prove that there exists a point R such that

$$RA_1 + RA_2 + \cdots + RA_n < s.$$

Solution: We claim that R can be any point between P and Q. Let k be the positive number such that $PR = kQR$. Suppose A_i does not lie on PQ. Let B_i be the point of intersection of RA_i with the line through P parallel to QA_i, as shown in Figure 5.2.1. Then $\angle RPB_i = \angle RQA_i$ and $\angle RB_iP = \angle RA_iQ$ by the Alternate Angles Theorem. Hence triangles PRB_i and QRA_i are similar by the AA Theorem, so that $RB_i = kRA_i$

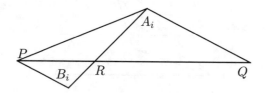

FIGURE 5.2.1

and $PB_i = kQA_i$. Applying the Triangle Inequality to PA_iB_i, we have

$$(1+k)RA_i = RA_i + RB_i = A_iB_i < PA_i + PB_i = PA_i + kQA_i.$$

Suppose A_i lies on PQ. Rotate A_i about R through a small angle and construct B_i as before. Then rotate both points about R back onto PQ. Now $A_iB_i \leq PA_i + PB_i$, so that

$$(1+k)RA_i \leq PA_i + kQA_i.$$

Since the A's are not all collinear, there is at least one that does not lie on PQ. It follows that

$$(1+k)\sum_{i=1}^{n} RA_i < \sum_{i=1}^{n} PA_i + k\sum_{i=1}^{n} QA_i = (1+k)s.$$

Hence $RA_1 + RA_2 + \cdots + RA_n < s$.

Remark: We could have simply taken $k = 1$ in the above solution and dealt with congruent triangles instead of similar triangles.

Problem 1930.3.

Inside an acute triangle ABC is a point P that is not the circumcenter. Prove that among the segments AP, BP and CP, at least one is longer and at least one is shorter than the circumradius of ABC.

First Solution: Let O be the circumcenter of triangle ABC and let R be its circumradius. For any point $P \neq O$ inside ABC, draw the perpendicular bisector of OP, as shown in Figure 5.2.2. It must intersect the perimeter of ABC twice, and at most once at a vertex. Hence it must intersect one of the sides, say AB, with A on the same side of it as P and B on the same side as O. Then $PA < OA = R$ and $PB > OB = R$.

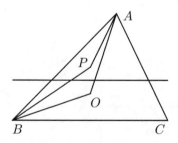

FIGURE 5.2.2

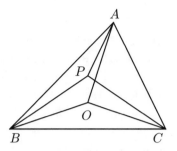

FIGURE 5.2.3

Second Solution: Let O be the circumcenter of triangle ABC and let R be its circumradius. For any point $P \neq O$ inside ABC, P must lie in one of triangles OBC, OCA and OAB, say the last. Then O must lie in one of triangles PCB or PAC, say the first. See Figure 5.2.3. By the Sss Inequality, $PB + PC > OB + OC = 2R$, so that either PB or PC must be greater than R. Similarly, $PA + PB < OA + OB = 2R$, so that either PA or PB must be shorter than R.

Problem 1943.2.

Let P be any point inside an acute triangle. Let D and d be respectively the maximum and minimum distances from P to any point on the perimeter of the triangle.

(a) Prove that $D \geq 2d$.

(b) Determine when equality holds.

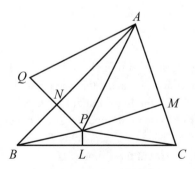

FIGURE 5.2.4

First Solution:

(a) Drop perpendiculars PL, PM and PN from P to BC, CA and AB respectively. By the Mean Value Principle, at least one of the six angles meeting at P is not less than $60°$. We may assume that $\angle APN \geq 60°$. Reflect P across AB to Q. Then $\angle AQN = \angle APN \geq 60°$ so that $\angle PAQ \leq 60°$. By the Angle-Side Inequality, $D \geq PA \geq PQ = 2PN \geq 2d$.

(b) In order for $D = 2d$ to hold, each of the six angles at P must be equal to $60°$. Then each of PA, PB and PC divides an angle of triangle ABC into two $30°$ angles. It follows that ABC is an equilateral triangle, and P is its center. On the other hand, if ABC is equilateral and P is its center, we certainly have $D = 2d$.

Second Solution:

(a) Let R and r be respectively the circumradius and inradius of ABC. By Euler's Inequality, $R \geq 2r$. By the preceding **Problem 1930.3**, we have $D \geq \max\{PA, PB, PC\} \geq R$. We now use a similar argument to prove that $r \geq d$. If P coincides with the incenter I of triangle ABC, we have $r = d$. Suppose $P \neq I$. Then P must be in one of the triangles IBC, ICA and IAB, say the last. Then the distance from P to AB is less than r, so that $r > d$.

(b) By (a), $D \geq R \geq 2r \geq 2d$. In order for $D = 2d$ to hold, we must have $D = R$ and $d = r$, so that P must simultaneously be the circumcenter and the incenter of triangle ABC. This implies that ABC is an equilateral triangle, with P as its center. On the other hand, if ABC is equilateral and P is its center, we certainly have $D = 2d$.

Third Solution:

(a) Through the centroid G of triangle ABC, draw the line parallel to BC cutting AB at Q and AC at R, the line parallel to CA cutting BC at S and BA at T, and the line parallel to AB cutting CA at U and CB at V. See the first diagram in Figure 5.2.5. Now any point inside ABC must lie in at least one of the trapezoids $BCRQ$, $CATS$ and $ABVU$, say the first. Let QR cut the altitude AK at L, as shown in the second diagram in Figure 5.2.5. Since G trisects the medians, $AL = 2LK$. Now drop the perpendicular PM from P to BC, and let QR cut AP at

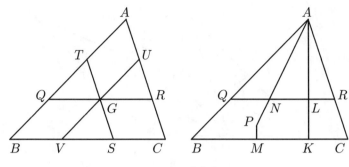

FIGURE 5.2.5

N. Applying the Angle-Side Inequality to triangle LAN, $AN \geq AL$. Hence $D \geq AP \geq AN \geq AL = 2LK \geq 2PM \geq 2d$.

(b) We may continue to assume that P is in $BCRQ$. In order for $D = 2d$ to hold, we must have $AP = AN = AL$. This means that P must lie on QR and AK. If P is in triangle GSV, then we must have $P = G$ since G is the only point in GSV which is on QR. If P is in $GSCR$ or $GVBQ$, then it must lie on ST or UV respectively, and we must still have $P = G$. Recall that $G = P$ must lie on AK. By symmetry, it must also lie on the other two altitudes. Hence the centroid of ABC is also its orthocenter, so that ABC is an equilateral triangle, with P as its center. On the other hand, if ABC is equilateral and P is its center, we certainly have $D = 2d$.

Remark: The result $D \geq 2d$ is true for any point inside any triangle, since the Third Solution did not make use of the condition that ABC is acute.

Problem 1940.3.

(a) Prove that for any triangle H_1, there exists a triangle H_2 whose side lengths are equal to the lengths of the medians of H_1.

(b) If H_3 is the triangle whose side lengths are equal to the lengths of the medians of H_2, prove that H_1 and H_3 are similar.

First Solution:

(a) Let AD, BE and CF be the medians of triangle $H_1 = ABC$. Rotate $180°$ about E so that A and C land on each other while the new positions of B, D and F are denoted by B', D' and F' respectively.

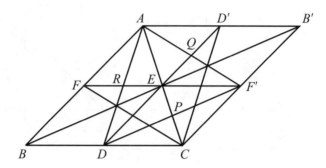

FIGURE 5.2.6

See Figure 5.2.6. Note that $ABCB'$ is a parallelogram. We have $AF = \frac{1}{2}AB = \frac{1}{2}CB' = CF'$, so that $AFCF'$ is also a parallelogram. Hence $AF' = CF$. Applying the Midpoint Theorem to triangle CBB', we have $DF' = \frac{1}{2}BB' = BE$. Hence $H_2 = ADF'$ is a triangle whose side lengths are equal respectively to the medians of triangle ABC.

(b) Let $H_1 = ABC$ and $H_2 = ADF'$ be as in (a). Note that E lies on DD' and FF'. Let AC cut DF' at P, AF' cut DD' at Q and AD cut FF' at R. Since $EDCF'$ is a parallelogram, $DP = PF'$ and $EP = PC$. Hence AP is a median of triangle ADF' and $AP = AE + EP = \frac{3}{4}AC$. Similarly, $EF'D'A$ and $EAFD$ are parallelograms. Hence DQ and $F'R$ are the other medians of triangle ADF'. We have $DQ = \frac{3}{4}DD' = \frac{3}{4}AB$ and $F'R = \frac{3}{4}FF' = \frac{3}{4}BC$. It follows that H_3 will have side-lengths proportional to the sides of H_1, so that the two triangles are similar by the sss Theorem.

Remark: For another solution to this problem, see page 114 in Chapter 6.

Problem 1936.2.

S is a point inside triangle ABC such that the areas of the triangles ABS, BCS and CAS are all equal. Prove that S is the centroid of ABC.

First Solution: Suppose S is the centroid of triangle ABC. Then the extension of AS cuts BC at its midpoint D, as shown in Figure 5.2.7. We have $[ABS] = [BAD] - [BSD] = [CAD] - [CSD] = [CAS]$. We also have $AS = 2SD$, so that $[BCS] = 2[BSD] = [ABS]$. For any point S' inside ABC other than its centroid S, S' is in one of triangles ABS,

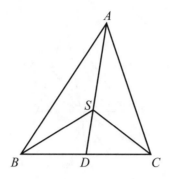

FIGURE 5.2.7

BCS and CAS. By symmetry, we may assume that S' is in CAS. Then $[CAS'] < [CAS] = \frac{1}{3}[ABC]$, and we cannot have $[ABS'] = [BCS'] = [CAS']$.

Second Solution: Extend AS to cut BC at D, as shown in Figure 5.2.7. Let k be the positive number such that $BD = kCD$. Then $[BSD] = k[CSD]$ and $[BAD] = k[CAD]$. It follows that

$$[ABS] = [BAD] - [BSD] = k([CAD] - [CSD]) = k[CAS].$$

Since $[ABS] = [ACS]$, we have $k = 1$ and AS is a median of triangle ABC. Similarly, so is BS, and S is indeed the centroid of ABC.

Third Solution: If a line that is equidistant from two points is not parallel to the line joining those two points, it must pass through the midpoint of the segment joining them. From $[CAS] = [ABS]$, the line AS is equidistant from B and C. Since S is inside triangle ABC, AS cannot be parallel to BC. It follows that S lies on the line joining A to the midpoint of BC, that is, AS is a median of triangle ABC. Similarly, so is BS, and S is indeed the centroid of ABC.

Fourth Solution: Since $[ABS] = \frac{1}{3}[ABC]$ and S is inside triangle ABC, it lies on a line parallel to AB and at a distance from AB one-third that from C to AB. Similarly, it lies on a line parallel to BC and at a distance from BC one-third that from A to BC. It follows that S is the point of intersection of these two lines. Let the extensions of AS and BS cut the opposite sides at D and E respectively, as shown in Figure 5.2.8. Then $SD = \frac{AD}{3}$ and $SE = \frac{BE}{3}$. By the Median Trisection Theorem, S is the centroid of triangle ABC.

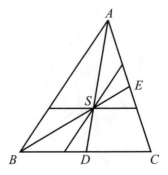

FIGURE 5.2.8

Problem 1942.3.

Let A', B' and C' be points on the sides BC, CA and AB, respectively, of an equilateral triangle ABC. If $AC' = 2C'B$, $BA' = 2A'C$ and $CB' = 2B'A$, prove that the lines AA', BB' and CC' enclose a triangle whose area is $\frac{1}{7}$ that of ABC.

First Solution: Start with an equilateral triangle DEF. Extend EF to A, FD to B and DE to C, with AF, BD and CE all equal to the side-length of DEF, as shown in Figure 5.2.9. Then ABC is also an equilateral triangle. Let AB cut the extension of CD at C', and the line through F parallel to DE at P. Applying the Midpoint Theorem to triangles AEC' and BFP, we have $AP = PC' = C'B$ so that $AC' = 2C'B$. Similarly, if the extension of AE cuts BC at A', and the extension of BF cuts CA at B', we will have $BA' = 2A'C$ and $CB' = 2B'A$. Precisely the same diagram will be obtained if we start with the equilateral triangle ABC. From $BD = DF$, we have $[BED] = [DEF]$, and from $DE = EC$,

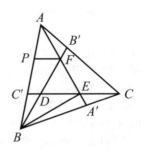

FIGURE 5.2.9

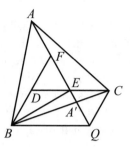

FIGURE 5.2.10

we have $[BEC] = [BED]$. It follows that $[BCD] = 2[DEF]$. Similarly, $[CAE] = [BAF] = 2[DEF]$ also. Hence $[DEF] = \frac{1}{7}[ABC]$.

Second Solution: Let AA' cut BB' at F and CC' at E, while BB' and CC' meet at D, as shown in Figure 5.2.9. By symmetry, DEF is an equilateral triangle, and $AF = BD = CE$. Extend AE to Q where QC is parallel to DF, as shown in Figure 5.2.10. Then CEQ is also an equilateral triangle. It follows that $CQ = CE = BD$. Hence $BQCD$ is a parallelogram, and we have $CD = BQ$. Now triangles $A'EC$ and $A'QB$ are similar by the AA Theorem. Since $BA' = 2A'C$, we have $CD = BQ = 2EC$. Hence E is the midpoint of CD. By symmetry, F is the midpoint of AE and D is the midpoint of BF. As in the First Solution, we have $[DEF] = \frac{1}{7}[ABC]$.

Third Solution: Draw lines parallel to BB' through C, A and the midpoint of $B'C$, and lines parallel to CC' through A, B and the midpoint of $C'A$, as shown in the first diagram of Figure 5.2.11. This produces a grid of nine parallelograms. Now draw the diagonal ℓ through A of the large

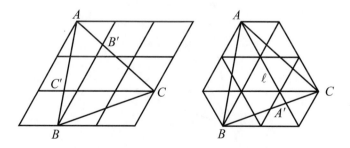

FIGURE 5.2.11

parallelogram along with diagonals of the small parallelograms parallel to ℓ. We have four equidistant parallel lines, the two outside ones passing through B and C, as shown in the second diagram of Figure 5.2.11. Thus they divide BC into three equal parts by the Intercept Theorem, so that A' lies on ℓ. Now triangle ABC is covered by 13 small triangles. The one determined by AA', BB' and CC' lies entirely within. The other 12 form three parallelograms, and exactly half of each is inside ABC. It follows that the area of each small triangle is one-seventh that of ABC, which is the desired result.

Fourth Solution: Let AA' cut BB' at F and CC' at E, while BB' and CC' meet at D, as shown in Figure 5.2.9. We may take $[ABC] = 21$. We have $[AC'D] = 2[BC'D]$ since $AC' = 2C'B$. Similarly, $[B'CD] = 2[B'AD]$ and $[B'CB] = 2[B'AB]$. In Figure 5.2.12, we have

$$[BCD] = [B'CB] - [B'CD]$$
$$= 2([B'AB] - [B'AD])$$
$$= 2[BAD]$$
$$= 2([AC'D] + [BC'D]$$
$$= 6[BC'D].$$

It follows that $[BCC'] = 7[BC'D]$. However, $[BCC'] = \frac{1}{3}[ABC] = 7$, so that $[BC'D] = 1$. By symmetry, $[AB'F] = [CA'E] = 1$ also, and we have

$$[DEF] = [ABC] - [BCC'] - [CAA'] - [ABB']$$
$$+[BC'D] + [CA'E] + [AB'F]$$
$$= 21 - 7 - 7 - 7 + 1 + 1 + 1 = 3.$$

It follows that $[DEF] = \frac{1}{7}[ABC]$.

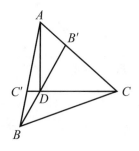

FIGURE 5.2.12

Remark: The Third and the Fourth Solutions show that the result is still true even if ABC is an arbitrary triangle, as neither makes use of the fact that ABC is equilateral. For another solution to this problem, see page 107 in Chapter 6.

Problem 1929.3.

Let p, q and r be three concurrent lines in the plane such that the angle between any two of them is $60°$. Let a, b and c be real numbers such that $0 < a \le b \le c$.

(a) Prove that the set of points whose distances from p, q and r are respectively less than a, b and c consists of the interior of a hexagon if and only if $a + b > c$.

(b) Determine the length of the perimeter of this hexagon when $a+b > c$.

Solution:

(a) Consider the points within distance a of p. They form a strip of width $2a$ with p as its centerline. Those points which are also within distance b of q form a parallelogram $ABCD$, as shown in Figure 5.2.13. In order for those points that are also within distance c of r to form a hexagon, the distance from A to r must be strictly greater than c. Let

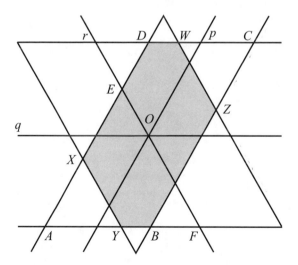

FIGURE 5.2.13

r cut AD at E and the extension of AB at F. Then AEF is an equilateral triangle. The sum of the distances from the point O to its three sides is $a + b$, where O is the point of concurrency of p, q and r. It follows from the Equilateral Triangle Altitude Theorem that the distance from A to EF is $a + b$, and the desired condition is indeed $a + b > c$.

(b) We have $AB = \frac{4a}{\sqrt{3}}$ and $AD = \frac{4b}{\sqrt{3}}$. The lines parallel to r and at a distance c from it will cut off from $ABCD$ the equilateral triangles AXY and CWZ. The altitude of AXY is $a + b - c$, so that $XY = AX = AY = WZ = CW = CZ = \frac{2(a+b-c)}{\sqrt{3}}$. Hence the perimeter of the hexagon is

$$2(AB + AD) - (AX + AY + CW + CZ) + (XY + WZ)$$

$$= \frac{4}{\sqrt{3}} \left[2a + 2b - (a + b - c) \right]$$

$$= \frac{4(a + b + c)}{\sqrt{3}}.$$

6
Geometry Problems
Part II

Problem 1932.2.

In triangle ABC, $AB \neq AC$. Let AF, AP and AT be the median, interior angle bisector and altitude from vertex A, with F, P and T on BC or its extension.

(a) Prove that P always lies between F and T.

(b) Prove that $\angle FAP < \angle PAT$ if ABC is an acute triangle.

Problem 1933.3.

The circles k_1 and k_2 are tangent at the point P. A line is drawn through P, cutting k_1 at A_1 and k_2 at A_2. A second line is drawn through P, cutting k_1 at B_1 and k_2 at B_2. Prove that the triangles PA_1B_1 and PA_2B_2 are similar.

Problem 1939.3.

ABC is an acute triangle. Three semicircles are constructed outwardly on the sides BC, CA and AB respectively. Construct points A', B' and C' on these semicircles respectively so that $AB' = AC'$, $BC' = BA'$ and $CA' = CB'$.

Problem 1941.3.

The hexagon $ABCDEF$ is inscribed in a circle. The sides AB, CD and EF are all equal in length to the radius. Prove that the midpoints of the other three sides determine an equilateral triangle.

Problem 1934.2.

Which polygon inscribed in a given circle has the property that the sum of the squares of the lengths of its sides is maximum?

Problem 1941.2.

Prove that if all four vertices of a parallelogram are lattice points and there are some other lattice points in or on the parallelogram, then its area exceeds 1.

Problem 1932.3.

Let α, β and γ be the interior angles of an acute triangle. Prove that if $\alpha < \beta < \gamma$, then $\sin 2\alpha > \sin 2\beta > \sin 2\gamma$.

Problem 1938.3.

Prove that for any acute triangle, there is a point in space such that every line segment from a vertex of the triangle to a point on the line joining the other two vertices subtends a right angle at this point.

Problem 1937.2.

Two circles in space are said to be tangent to each other if they have a common tangent at the same point of tangency. Assume that there are three circles in space that are mutually tangent at three distinct points. Prove that they either all lie in a plane or all lie on a sphere.

6.1   Discussion

6.1.1   Circles

In the preceding Chapter, we have dealt with the geometry of triangles and quadrilaterals. In this subsection, we turn our attention to the geometry of circles. In the remaining subsections, we consider other approaches which are not purely synthetic, and give a brief introduction to solid geometry.

In Chapter 5, we have defined a circle as the set of all points at a fixed distance from a fixed point called its center. An **arc** of a circle is any connected part of it. If the arc is exactly one half of the circle, then it is called a **semicircle**. An arc of a circle that is less than a semicircle is called a **minor arc**, while one greater than a semicircle is called a **major arc**. Unless otherwise stated, an arc is taken to be a minor arc.

A line cuts a circle in at most two points. If it cuts the circle in exactly two points, it is called a **secant**, and the part of the secant inside the circle is called a **chord**. If the chord passes through the center of the circle, then it is called a **diameter**, and each half of a diameter is called a **radius**. The terms *diameter* and *radius* are also used to denote their lengths. Thus the radius is the fixed distance from any point on the circle to its center. The context should make it clear which interpretation is intended. If the line intersects a circle in exactly one point, then it is called a **tangent**, and the point is called the **point of tangency**.

Here is another geometric bargain associated with a chord of a circle.

Chord-Radius Theorem.   *Consider the following three statements about a line, a circle and a chord of the circle.*
(1) *The line passes through the center of the circle.*
(2) *The line passes through the midpoint of the chord.*
(3) *The line is perpendicular to the chord.*
If any two of these statements hold, then the third one is also true.

Proof.   This result is an easy consequence of the properties of the perpendicular bisector of the chord in question.

Tangent-Radius Theorem.   *Let P be a point on a circle with center O. A line passing through P is a tangent of the circle if and only if it is perpendicular to OP.*

Proof.   Suppose the line is indeed a tangent. Since every point other than P on it is outside the circle, OP is the shortest distance from O to this line. It follows that OP is perpendicular to the tangent. Conversely, suppose the line is perpendicular to the radius of the circle at that point. If the line cuts the circle at a second point Q, then OPQ is a right triangle with OQ as the hypotenuse, so that $OQ > OP$. This contradicts $OP = OQ$.

Intersecting Tangents Theorem.   *From a point P outside a circle, tangents are drawn to the circle at the points S and T. Then $PS = PT$.*

Proof.   Let O be the center of the circle. Then $OP = OP$ and $OS = OT$. By the Tangent-Radius Theorem, $\angle OSP = 90° = \angle OTP$. Hence triangles POS and POT are congruent, so that $PS = PT$.

Let A and B be two points on a circle with center O that are not diametrically opposite. Then $\angle AOB$ is called the **angle subtended at the center** by the arc AB. For any point C on the major arc AB, $\angle ACB$ is called the **angle subtended at the circle** by the arc AB. If AB is a diameter and C is any point on one semicircle defined by AB, then $\angle ACB$

is called the **angle subtended at a semicircle** by the other semicircle defined by AB.

It turns out that the angle subtended at the circle, or at a semicircle, is independent of the specific point C. This follows from the following very important result.

Thales' Theorem. *For any arc, the angle subtended at the center is double the angle subtended at the circle.*

FIGURE 6.1.1

Proof. In the first diagram of Figure 6.1.1, we have $\angle OAC = \angle OCA$ since $OA = OC$. Hence $\angle AOB = \angle OAC + \angle OCA = 2\angle ACB$ by the Exterior Angle Theorem. In the second and third diagrams of Figure 6.1.1, draw the diameter CD. Then

$$\angle AOB = \angle BOD \pm \angle AOD = 2\angle BCD \pm 2\angle ACD = 2\angle ACB.$$

If C is on the minor arc AB, then $\angle ACB = \frac{1}{2}(360° - \angle AOB)$. This can be proved in a similar way. Thales' Theorem has many important corollaries.

Semicircle-Angle Theorem. *If AB is a diameter of a circle, then $\angle ACB = 90°$ if and only if C lies on the circle.*

Proof. Suppose C lies on the circle. Then

$$\angle ACB = \angle ACO + \angle BCO = \angle CAO + \angle CBO = 180° - \angle ACB.$$

Hence $\angle ACB = 90°$. Suppose C is inside the circle. Extend AC to cut the circle again at D. Then $\angle ACB > \angle ADB = 90°$. Finally, if C is outside the circle, let AC cut the circle again at D. Then we have $\angle ACB < \angle ADB = 90°$.

A quadrilateral whose vertices lie on a circle is said to be **cyclic**, and its vertices are said to be **concyclic**.

Circle-Angle Theorem. *ABCD is a cyclic quadrilateral if and only if* $\angle ACB = \angle ADB$.

Proof. Suppose $ABCD$ is cyclic. By Thales' Theorem, each of $\angle ACB$ and $\angle ADB$ is one half of $\angle AOB$, where O is the center of the circle. Conversely, suppose $\angle ACB = \angle ABD$ but D is inside the circumcircle of triangle ABC. Let the extension of AD cut the circle again at E. Then we have $\angle AEB = \angle ACB$. By the Exterior Angle Inequality, $\angle AEB < \angle ADB = \angle ACB$, which is a contradiction. The case where D is outside the circle leads to a contradiction in a similar manner.

The technique used in proving the converse of the preceding result, by assuming that a point is not where it is supposed to be and then deriving a contradiction, is often called the *method of false position*. It is a basic and useful tool. When the proof of the converse of a theorem is omitted, it usually means that it can be handled by this method.

Opposite Angles Theorem. *ABCD is a cyclic quadrilateral if and only if* $\angle ABC + \angle CDA = 180°$.

Proof. Suppose $ABCD$ is cyclic. By Thales' Theorem, one of $\angle ABC$ and $\angle ADC$ is one-half that of $\angle AOC$, and the other is one-half that of the reflex angle AOC. It follows that their sum is $180°$.

Tangent-Angle Theorem. Let PT be a tangent to a circle at the point T. Let A be any point on the circle and C be any point on the circle on the opposite side to P of the line AT. Then $\angle PTA = \angle ACT$.

Proof. Let BT be a diameter of the circle. Suppose B is on the opposite side to P of the line AT, as shown in the first diagram of Figure 6.1.2. Now $\angle ABT + \angle ATB = 180° - \angle BAT = 90°$ by the Semicircle-Angle Theorem and $\angle ATP + \angle ATB = 90°$ by the Tangent-Radius Theorem. It follows that $\angle ATP = \angle ABT$. By the Circle-Angle Theorem, $\angle ACT =$

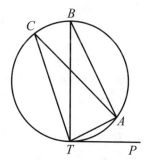 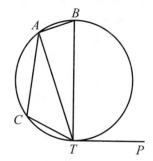

FIGURE 6.1.2

$\angle ABT = \angle ATP$. Suppose instead B is on the same side as P of the line AT, as shown in the second diagram of Figure 6.1.2. By the Semicircle-Angle Theorem,

$$\angle ABT + \angle ATB = 180° - \angle BAT = 90°.$$

Now $\angle ATP - \angle ATB = 90°$ by the Tangent-Radius Theorem. Hence $\angle ATP = 180° - \angle ABT$. By the Opposite Angles Theorem,

$$\angle ACT = 180° - \angle ABT = \angle ATP.$$

The **power** of a point P with respect to a circle with center O and radius r is defined to be the quantity $OP^2 - r^2$. The power is negative if P is inside the circle, zero if it is on the circle and positive if it is outside the circle. From a point P outside the circle, draw a tangent to the circle, touching it at the point T. By the Tangent-Radius Theorem and Pythagoras' Theorem, $PT^2 = OP^2 - r^2$, so that the length of the tangent PT is equal to the square root of the power of P.

Intersecting Chords Theorem.   *Let AB and CD be any two intersecting segments and let P be the point of intersection. Then $PA \cdot PB = PC \cdot PD$ if and only if A, B, C and D are concyclic.*

Proof.   Suppose $ACBD$ is inscribed in a circle with center O and radius r. Let M be the midpoint of AB with P on the segment AM. Then

$$PA \cdot PB = (AM - PM)(BM + PM) = AM^2 - PM^2$$
$$= (AM^2 + OM^2) - (PM^2 + OM^2)$$
$$= r^2 - OP^2,$$

which is the negative of the power of P. This is independent of the choice of AB. It follows that $PA \cdot PB = PC \cdot PD$.

Intersecting Secants Theorem.   *Let AB and CD be any two segments whose extensions intersect at a point P. Then $PA \cdot PB = PC \cdot PD$ if and only if A, B, C and D are concyclic.*

Proof.   First, suppose that $ABDC$ is inscribed in a circle with center O and radius r. Let A be between P and B, and let M be the midpoint of AB. Then

$$PA \cdot PB = (PM - AM)(PM + BM) = PM^2 - BM^2$$
$$= (PM^2 + OM^2) - (BM^2 + OM^2)$$
$$= OP^2 - r^2,$$

which is the power of P. This is independent of the choice of AB. It follows that $PA \cdot PB = PC \cdot PD$.

Secant-Tangent Theorem. *Let A be a point on the segment PB and T be a point on a circle through A and B. Then $PA \cdot PB = PT^2$ if and only if PT is a tangent to the circle.*

Proof. First, suppose that PT is tangent to the circumcircle of triangle BAT. Then both $PA \cdot PB$ and PT^2 are equal to the power of P.

Suppose P is a point having equal powers with respect to the circle having center O_1 and radius r_1, and the circle having center O_2 and radius r_2. Then $O_1P^2 - O_2P^2 = r_1^2 - r_2^2$. Drop the perpendicular PQ from P onto O_1O_2. Then

$$O_1Q^2 - O_2Q^2 = (O_1P^2 - PQ^2) - (O_2P^2 - PQ)^2$$
$$= O_1P^2 - O_2P^2.$$

It follows that Q also has equal powers with respect to these two circles. In fact, any point on the line PQ has this property, but no other points do. Hence the set of points with equal powers with respect to two circles constitute a line perpendicular to the line joining the two centers.

This line is called the **radical axis** of the two circles. If they intersect each other in two points, their radical axis is their common chord. If they touch each other at one point, then their radical axis is their common tangent at that point. The tangents to two circles from any exterior points on their radical axis have the same length.

6.1.2 Coordinate Geometry

Coordinate Geometry, or Analytic Geometry, uses an algebraic approach to tackle geometric problems. Since this is a standard topic in the North American curriculum, we will be brief and select only some topics not usually covered in the classroom.

A horizontal line and a vertical line are chosen in the plane. The horizontal line is called the **x-axis** and the vertical line the **y-axis**. Their point of intersection is called the **origin**. For an arbitrary point in the plane, its **x-coordinate** is its distance from the y-axis, which is taken to be positive if the point is to the right of the y-axis and negative if it is to the left. If the point is on the y-axis, then it is 0. The **y-coordinate** of a point is its distance from the x-axis, which is taken to be positive if the point is above the x-axis and negative if it is below. If the point is on the x-axis, then it is 0. Thus each point can be identified with a pair of real numbers.

Every two distinct points determine a unique straight line. Let the coordinates of the points be (a, b) and (c, d). The **slope** m of the line joining them is the ratio of the change in the y-coordinates to the change in the x-coordinates. In order words, $m = \frac{b-d}{a-c}$. If (x, y) is an arbitrary point of this line, it must determine the same slope m with either of the two points. Hence (x, y) satisfies $\frac{y-b}{x-a} = \frac{b-d}{a-c}$. This is a linear equation in two variables. Conversely, such an equation always represents a line.

Higher order equations represent more complicated curves. Since linear equations are easy to solve, it may be advantageous to impose on a geometric problem a coordinate system if the problem involves primarily straight lines. However, when circles are involved, especially mutually tangent ones, the algebra can become very cumbersome.

A useful result is the **Coordinates-Area Formula**, which gives the area of a triangle in terms of the coordinates of its vertices. If these coordinates are (x_1, y_1), (x_2, y_2) and (x_3, y_3), then the area of the triangle is given by $\frac{1}{2}|x_1(y_2 - y_3) + x_2(y_3 - y_1) + x_3(y_1 - y_2)|$. Note that the expression is symmetric in the indices, so that it does not matter which vertex has which coordinates. The absolute value sign is there to ensure that the area is nonnegative.

We shall prove this formula by packing the triangle into a rectangle with sides parallel to the axes. Five different configurations may arise as shown in Figure 6.1.3. We consider only the fourth case.

Let the coordinates of A be (x_1, y_1), those of B (x_2, y_2) and those of C (x_3, y_3) as shown in Figure 6.1.4. Drop the perpendiculars BE and BF from B onto AD and CD respectively. Then we have

$$[CAD] = \frac{1}{2}(x_3 - x_1)(y_3 - y_1), \quad [ABE] = \frac{1}{2}(x_2 - x_1)(y_2 - y_1),$$

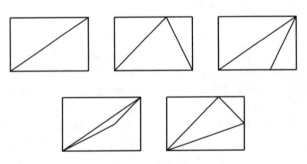

FIGURE 6.1.3

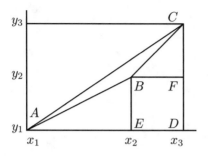

FIGURE 6.1.4

$$[CBF] = \frac{1}{2}(x_3 - x_2)(y_3 - y_2) \quad \text{and} \quad [BEDF] = (x_3 - x_2)(y_2 - y_1).$$

Hence

$$[ABC] = [CAD] - [ABE] - [CBF] - [BEDF],$$

and algebraic computations yield the desired result.

We now give an application of this result.

Problem 1942.3.

Let A', B' and C' be points on the sides BC, CA and AB, respectively, of an equilateral triangle ABC such that $AC' = 2C'B$, $BA' = 2A'C$ and $CB' = 2B'A$. Prove that the lines AA', BB' and CC' enclose a triangle whose area is $\frac{1}{7}$ that of ABC.

Fifth Solution: In the coordinate plane, let A be the point $(0,0)$, B be $(3, 3\sqrt{3})$, C be $(6,0)$, A' be $(5, \sqrt{3})$, B' be $(2,0)$ and C' be $(2, 2\sqrt{3})$. Then ABC is indeed an equilateral triangle and its sides are divided in the given manner. Since the side-length of ABC is 6, its area is $9\sqrt{3}$. Note that the equations of the lines AA', BB' and CC' are respectively $y = \frac{\sqrt{3}}{5}x$, $y = 3\sqrt{3}x - 6\sqrt{3}$ and $y = -\frac{\sqrt{3}}{2}x + 3\sqrt{3}$. Solving these equations in pairs, we find that the point D of intersection of BB' and CC' is $(\frac{18}{7}, \frac{12\sqrt{3}}{7})$, the point E of intersection of CC' and AA' is $(\frac{30}{7}, \frac{6\sqrt{3}}{7})$ and the point F of intersection of AA' and BB' is $(\frac{15}{7}, \frac{3\sqrt{3}}{7})$. By the Coordinates-Area Formula, the area of DEF is given by

$$\frac{1}{2}\left| \frac{18}{7}\left(\frac{6\sqrt{3}}{7} - \frac{3\sqrt{3}}{7}\right) + \frac{30}{7}\left(\frac{3\sqrt{3}}{7} - \frac{12\sqrt{3}}{7}\right) + \frac{15}{7}\left(\frac{12\sqrt{3}}{7} - \frac{6\sqrt{3}}{7}\right)\right| = \frac{9\sqrt{3}}{7}.$$

This is exactly one-seventh the area of ABC.

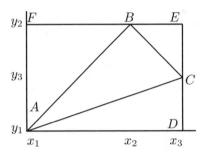

FIGURE 6.1.5

The points whose coordinates are both integers are called **lattice points**, and a polygon whose vertices are all lattice points is called a lattice polygon. **Pick's Theorem** states that the area of a lattice polygon is given by $I + \frac{1}{2}O - 1$, where I is the number of lattice points inside the polygon and O is the number of lattice points on its boundary.

We shall prove this result by mathematical induction. As basis, consider a triangle. We have seen in Figure 6.1.3 that there are essentially five ways in which a triangle can be packed inside a rectangle with sides parallel to the axes. We shall treat only the fifth case here, as shown in Figure 6.1.5.

Let the coordinates of A be (x_1, y_1), those of B (x_2, y_2) and those of C (x_3, y_3). Let there be a lattice points on BC, b on CA and c on AB, not counting the endpoints. Then $B = a + b + c + 3$. The number of lattice points inside the rectangle $ADEF$ is given by $N = (x_3 - x_1 - 1)(y_2 - y_1 - 1)$. The number of lattice points inside triangle BCE is $N_1 = \frac{1}{2}\big[(y_2 - y_3 - 1)(x_3 - x_2 - 1) - a\big]$, that inside ACD is $N_2 = \frac{1}{2}\big[(y_3 - y_1 - 1)(x_3 - x_1 - 1) - b\big]$ and that inside FAB is $N_3 = \frac{1}{2}\big[(y_2 - y_1 - 1)(x_2 - x_1 - 1) - c\big]$. Using the fact that I is given by $N - N_1 - N_2 - N_3 - a - b - c$, we can show that $I + \frac{1}{2}O - 1$ yields the same result as that given by the Coordinates-Area Formula.

Suppose Pick's Theorem has been proved for all lattice polygons with up to n sides. Consider a lattice polygon with $n + 1$ sides. Divide it into two polygons by a diagonal that lies entirely inside the $(n+1)$-gon. Let d be the number of lattice points on this diagonal, not counting the endpoints. Let I_1 and O_1 be the numbers of interior and boundary lattice points of one of the two resulting polygons, and let I_2 and O_2 be that of the other. Then $I = I_1 + I_2 + d$ and $O = O_1 + O_2 - 2d - 2$. By the induction hypotheses, the area of the $(n+1)$-gon is $(I_1 + \frac{1}{2}O_1 - 1) + (I_2 + \frac{1}{2}O_2 - 1) = I + \frac{1}{2}O - 1$ as desired.

We now give an application of Pick's Formula.

Problem 1942.2.

Let a, b, c and d be integers such that for all integers m and n, there exist integers x and y such that $ax + by = m$ and $cx + dy = n$. Prove that $ad - bc = \pm 1$.

Second Solution: Partition the plane into parallelograms determined by the two families of lines $\{y = \frac{d}{b}(x - ka) + kc\}$ and $\{y = \frac{c}{a}(x - \ell b) + \ell d\}$, where k and ℓ are arbitrary integers. The vertices of these parallelograms are precisely the points of the form $(ax + by, cx + dy)$. The given condition shows that every lattice point (m, n) is a vertex of one of the parallelograms. The parallelogram with vertices $(0, 0)$, (a, c), (b, d) and $(a + b, c + d)$, therefore, contains no lattice points other than its vertices. By Pick's Theorem, its area is 1. By the Coordinates-Area Formula, the area of the triangle with vertices $(0, 0), (a, b)$ and (c, d) is $\frac{1}{2}|ad - bc| = \frac{1}{2}$. It follows that $ad - bc = \pm 1$.

6.1.3   Trigonometry

In similar triangles, the corresponding sides are in the same ratio. This allows us to measure angles by means of their **trigonometric ratios: sine, cosine, tangent, cotangent, secant** and **cosecant**. These are defined, along with their standard abbreviations, as follows.

Let O be the vertex of an angle θ. From any point B on one arm, drop a perpendicular BA onto the other arm, as shown in Figure 6.1.6. Then

$$\sin\theta = \frac{AB}{OB}, \quad \cos\theta = \frac{OA}{OB}, \quad \tan\theta = \frac{AB}{OA},$$

$$\csc\theta = \frac{OB}{AB}, \quad \sec\theta = \frac{OB}{OA}, \quad \cot\theta = \frac{OA}{AB}.$$

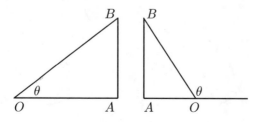

FIGURE 6.1.6

If θ is obtuse, OA is taken to be negative, so that only $\sin\theta$ and $\csc\theta$ are positive.

Most of the time, we work only with sines and cosines. This is because the other trigonometric ratios can be expressed in terms of them. We have $\tan\theta = \frac{\sin\theta}{\cos\theta}$, $\cot\theta = \frac{\cos\theta}{\sin\theta}$, $\sec\theta = \frac{1}{\cos\theta}$ and $\csc\theta = \frac{1}{\sin\theta}$.

Here are some very simple identities which follow directly from the definitions of the ratios: $\sin(90° - \theta) = \cos\theta$, $\cos(90° - \theta) = \sin\theta$, $\sin(180° - \theta) = \sin\theta$ and $\cos(180° - \theta) = -\cos\theta$. From Pythagoras' Theorem, we have $\sin^2\theta + \cos^2\theta = 1$.

Some useful values are $\sin 0° = \cos 90° = 0$, $\sin 30° = \cos 60° = \frac{1}{2}$, $\sin 45° = \cos 45° = \frac{1}{\sqrt{2}}$, $\sin 60° = \cos 30° = \frac{\sqrt{3}}{2}$ and $\sin 90° = \cos 0° = 1$. Note that as θ increases from $0°$ to $90°$, $\sin\theta$ increases while $\cos\theta$ decreases. As θ increases from $90°$ to $180°$, both $\sin\theta$ and $\cos\theta$ decrease.

The sine ratio can be used to express the area of a triangle. Suppose the triangle has two sides of lengths a and b, with an angle γ between them. If we take the side of length b as base, it is easy to see that the altitude on this base has length $a\sin\gamma$. Then its area is given by $\frac{1}{2}ab\sin\gamma$. This is known as the **Sine-Area Formula**.

Let R be the circumradius of triangle ABC with $BC = a, CA = b$, $AB = c$, $\angle CAB = \alpha$, $\angle ABC = \beta$ and $\angle BCA = \gamma$. Draw the diameter CD, as shown in Figure 6.1.7. Note that $\angle DBC = 90°$ and $BC = CD\sin CDB$. There are two possibilities. We have either $\angle CDB = \angle CAB = \alpha$ or $\angle CDB = 180° - \angle CAB = 180° - \alpha$. In both cases, $\sin CDB = \sin\alpha$ so that $a = 2R\sin\alpha$. It follows by symmetry that

$$\frac{1}{2R} = \frac{\sin\alpha}{a} = \frac{\sin\beta}{b} = \frac{\sin\gamma}{c}.$$

This is known as the **Law of Sines**.

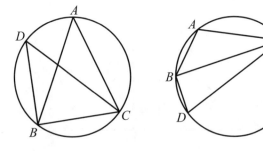

FIGURE 6.1.7

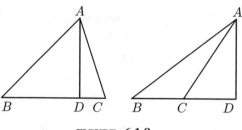

FIGURE 6.1.8

The **Law of Cosines** is a generalization of Pythagoras' Theorem. Let AD be the altitude from A to BC in triangle ABC, as shown in Figure 6.1.8. Then

$$AB^2 = AD^2 + (BC \mp CD)^2$$
$$= AD^2 + BC^2 + CD^2 \mp 2BC \cdot CD$$
$$= BC^2 + CA^2 - 2BC \cdot CA \cos BCA.$$

With the standard notations, we have

$$c^2 = a^2 + b^2 - 2ab \cos \gamma.$$

Note that when $\gamma = 90°$, the Law of Cosines reduces to $c^2 = a^2 + b^2$.

We now come to the important **Compound Angle Formulas**. We only prove the case where $\alpha + \beta$ is acute, as shown in Figure 6.1.9. The other case can be handled in a similar manner.

Let $\angle AOB = \alpha$ and $\angle BOC = \beta$. Let AB be perpendicular to OA, and let BC be perpendicular to OB. Draw a line through C parallel to

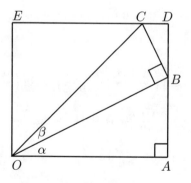

FIGURE 6.1.9

OA, cutting AB at D. Drop the perpendicular OE from O onto CD. Then $\angle CBD = \alpha$ and $\angle OCE = \alpha + \beta$. We have

$$\sin(\alpha + \beta) = \frac{OE}{OC}$$

$$= \frac{AB}{OC} + \frac{BD}{OC}$$

$$= \frac{AB}{OB}\frac{OB}{OC} + \frac{BD}{BC}\frac{BC}{OC}$$

$$= \sin \alpha \cos \beta + \cos \alpha \sin \beta.$$

We also have

$$\cos(\alpha + \beta) = \frac{CE}{OC}$$

$$= \frac{OA}{OC} - \frac{CD}{OC}$$

$$= \frac{OA}{OB}\frac{OB}{OC} - \frac{CD}{BC}\frac{BC}{OC}$$

$$= \cos \alpha \cos \beta - \sin \alpha \sin \beta.$$

We have $\sin \alpha = \sin(\alpha - \beta) \cos \beta + \cos(\alpha - \beta) \sin \beta$ as well as $\cos \alpha = \cos(\alpha - \beta) \cos \beta - \sin(\alpha - \beta) \sin \beta$. Subtracting $\sin \beta$ times the second from $\cos \beta$ times the first and using $\sin^2 \beta + \cos^2 \beta = 1$,

$$\sin(\alpha - \beta) = \sin \alpha \cos \beta - \cos \alpha \sin \beta.$$

Similarly, we can prove that

$$\cos(\alpha - \beta) = \cos \alpha \cos \beta + \sin \alpha \sin \beta.$$

We can express sums and differences of sines and cosines as products as follows:

$$\sin \alpha + \sin \beta = \sin \left(\frac{\alpha + \beta}{2} + \frac{\alpha - \beta}{2} \right) + \sin \left(\frac{\alpha + \beta}{2} - \frac{\alpha - \beta}{2} \right)$$

$$= \sin \frac{\alpha + \beta}{2} \cos \frac{\alpha - \beta}{2} + \cos \frac{\alpha + \beta}{2} \sin \frac{\alpha - \beta}{2}$$

$$+ \sin \frac{\alpha + \beta}{2} \cos \frac{\alpha - \beta}{2} - \cos \frac{\alpha + \beta}{2} \sin \frac{\alpha - \beta}{2}$$

$$= 2 \sin \frac{\alpha + \beta}{2} \cos \frac{\alpha - \beta}{2}.$$

Similarly, we can prove the following:

$$\sin \alpha - \sin \beta = 2 \cos \frac{\alpha + \beta}{2} \sin \frac{\alpha - \beta}{2},$$

$$\cos \alpha + \cos \beta = 2 \cos \frac{\alpha + \beta}{2} \cos \frac{\alpha - \beta}{2},$$

$$\cos \alpha - \cos \beta = -2 \sin \frac{\alpha + \beta}{2} \sin \frac{\alpha - \beta}{2}.$$

The Compound Angle Formulas give rise to the **Double Angle Formulas** $\sin 2\theta = 2 \sin \theta \cos \theta$ and $\cos 2\theta = 2 \cos^2 \theta - 1 = 1 - 2 \sin^2 \theta$. These in turn yield $\sin \theta \cos \theta = \frac{1}{2} \sin 2\theta$, $\sin^2 \theta = \frac{1}{2}(1 - \cos 2\theta)$ and $\cos^2 \theta = \frac{1}{2}(1 + \cos 2\theta)$.

6.1.4 Vectors and Complex Numbers

The concept of vectors is very useful, both in mathematics and physics. A **vector** is a directed segment. It is determined by its length and direction, but is independent of its actual position. The vector from the initial point A to the terminal point B is denoted by $\overline{AB}$. If we do not wish to specify the initial and terminal points, we may use the notation **v** to denote a vector. In particular, we use **0** to denote the vector whose initial point coincides with its terminal point. This vector has zero length and unspecified direction.

The operation of **vector addition** is defined as follows. If we want to add two vectors, place the initial point of the second vector at the terminal point of the first vector. The sum of these two vectors goes from the initial point of the first to the terminal point of the second. For instance, the sum of the vectors $\overline{AB}$ and $\overline{BC}$ is the vector $\overline{AC}$. Since $\overline{AC} + \overline{CA} = \mathbf{0}$, we have $\overline{AB} + \overline{BC} + \overline{CA} = \mathbf{0}$.

Complete the parallelogram $ABCD$ as shown in Figure 6.1.10. Then $\overline{BC} + \overline{AB} = \overline{AD} + \overline{DC} = \overline{AC}$, so that vector addition is *commutative*,

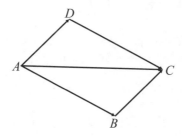

FIGURE 6.1.10

that is, $\mathbf{u} + \mathbf{v} = \mathbf{v} + \mathbf{u}$. It is easy to see that it is also *associative*. In other words, $(\mathbf{u} + \mathbf{v}) + \mathbf{w} = \mathbf{u} + (\mathbf{v} + \mathbf{w})$.

If we want to subtract one vector from another, make their initial points coincide. The difference is the vector going from the terminal point of the second vector to that of the first. Note for instance that $\overline{AC} - \overline{AD} = \overline{DC}$.

We now give several applications of the vector method.

Problem 1935.2.

Prove that a finite point set cannot have more than one center of symmetry.

Fourth Solution: Suppose O and O' are two centers of symmetry of a finite point set $S = \{P_1, P_2, \ldots, P_n\}$. For any P_i, let P_j be the point symmetric to P_i with respect to O. Then P_j also belongs to S and we have $\overline{OP_i} + \overline{OP_j} = \mathbf{0}$. It follows that $\sum_{i=1}^{n} \overline{OP_i} = \mathbf{0}$. Similarly, $\sum_{i=1}^{n} \overline{O'P_i} = \mathbf{0}$. Hence $n\overline{OO'} = \sum_{i=1}^{n} (\overline{OP_i} - \overline{O'P_i}) = \mathbf{0}$. It follows that $O = O'$.

The operation of **scalar multiplication** is defined as follows. If we multiply a vector $\mathbf{r}$ by a positive real number λ, we get a vector in the same direction as $\mathbf{r}$ but λ times as long. If λ is negative, the resulting vector is in the opposite direction and $|\lambda|$ times as long. This operation satisfies $\lambda(\mu\mathbf{v}) = (\lambda\mu)\mathbf{v}$, $(\lambda + \mu)\mathbf{v} = \lambda\mathbf{v} + \mu\mathbf{v}$ and $\lambda(\mathbf{u} + \mathbf{v}) = \lambda\mathbf{u} + \lambda\mathbf{v}$. The word scalar means a real number and is used to emphasize its difference from a vector. Note that subtraction may be considered as a special case of addition and scalar multiplication in that $\mathbf{u} - \mathbf{v} = \mathbf{u} + (-1)\mathbf{v}$.

Problem 1940.3

(a) Prove that for any triangle H_1, there exists a triangle H_2 whose side lengths are equal to the lengths of the medians of H_1.

(b) If H_3 is the triangle whose side lengths are equal to the lengths of the medians of H_2, prove that H_1 and H_3 are similar.

Second Solution:

(a) Let D, E and F be the respective midpoints of the sides BC, CA and AB of triangle $H_1 = ABC$. Then

$$\overline{AD} = \overline{AB} + \frac{1}{2}\overline{BC},$$

$$\overline{BE} = \overline{BC} + \frac{1}{2}\overline{CA},$$

$$\overline{CF} = \overline{CA} + \frac{1}{2}\overline{AB}.$$

Hence $\overline{AD}+\overline{BE}+\overline{CF} = \frac{3}{2}(\overline{AB}+\overline{BC}+\overline{CA}) = \mathbf{0}$. This means that if the vectors $\overline{AD}$, $\overline{BE}$ and $\overline{CF}$ are laid end to end, they will close up to form another triangle H_2.

(b) Let $H_1 = ABC$ be as in (a) and let H_2 be triangle PQR where $\overline{PQ} = \overline{AD}$, $\overline{QR} = \overline{BE}$ and $\overline{RP} = \overline{CF}$. Let X, Y and Z be the respective midpoints of QR, RP and PQ. Then

$$\overline{PX} = \overline{PQ} + \frac{1}{2}\overline{QR}$$

$$= \overline{AB} + \frac{1}{2}\overline{BC} + \frac{1}{2}(\overline{BC} + \frac{1}{2}\overline{CA})$$

$$= \overline{AB} + \overline{BC} + \overline{CA} - \frac{3}{4}\overline{CA}$$

$$= \frac{3}{4}\overline{AC}.$$

Similarly, $\overline{QY} = \frac{3}{4}\overline{BA}$ and $\overline{RZ} = \frac{3}{4}\overline{CB}$. Hence PX, QY and RZ will form a triangle H_3 with sides three-quarters the lengths of the sides of triangle ABC. It follows that H_3 is indeed similar to H_1.

The length of a vector $\mathbf{v}$ is denoted by $|\mathbf{v}|$. The **inner product** of two vectors $\mathbf{u}$ and $\mathbf{v}$, denoted by $\mathbf{u} \cdot \mathbf{v}$ or simply $\mathbf{uv}$, is defined as the scalar $|\mathbf{u}||\mathbf{v}| \cos\theta$ where θ is the angle between $\mathbf{u}$ and $\mathbf{v}$. Note that $\mathbf{uu} = |\mathbf{u}|^2$ while $\mathbf{uv} = 0$ if $\mathbf{u}$ and $\mathbf{v}$ are perpendicular vectors. This operation satisfies the following properties: $\mathbf{uv} = \mathbf{vu}$, $\mathbf{u}(\mathbf{v} + \mathbf{w}) = \mathbf{uv} + \mathbf{uw}$ and $\lambda\mathbf{u} \cdot \mu\mathbf{v} = (\lambda\mu)\mathbf{uv}$.

In triangle ABC, let $BC = a$, $CA = b$, $AB = c$ and $\angle BCA = \gamma$. Let $\overline{CB} = \mathbf{a}$, $\overline{CA} = \mathbf{b}$ and $\overline{AB} = \mathbf{c}$. Then $\mathbf{c} = \mathbf{a} - \mathbf{b}$, so that $c^2 = (\mathbf{a} - \mathbf{b})^2 = a^2 + b^2 - 2ab\cos\gamma$. Thus we have an alternative proof of the Law of Cosines.

We can combine the vector approach with the coordinate approach as follows. A vector of length 1 is called a unit vector. Let $\mathbf{i}$ be the unit vector in the direction of the positive x-axis, and $\mathbf{j}$ be the unit vector in the direction of the positive y-axis. Then every vector $\mathbf{u}$ in the plane can be expressed in the form $a\mathbf{i} + b\mathbf{j}$, where (a, b) is the terminal point of the vector if we place its initial point at $(0, 0)$.

Note that $\mathbf{i}$ and $\mathbf{j}$ can be chosen to be any two perpendicular unit vectors. If the angle between $\mathbf{u}$ and $\mathbf{i}$ is θ, then $a = |\mathbf{u}| \cos\theta = \mathbf{u} \cdot \mathbf{i}$ while $b = |\mathbf{u}| \sin\theta = \mathbf{u} \cdot \mathbf{j}$. In other words, the projection of a vector $\mathbf{u}$ in the direction of a unit vector $\mathbf{i}$ is their inner product $\mathbf{u} \cdot \mathbf{i}$, and hence the vector $(\mathbf{u} \cdot \mathbf{i})\mathbf{i}$ is the *component* of $\mathbf{u}$ in the direction of $\mathbf{i}$.

If $\mathbf{v} = c\mathbf{i} + d\mathbf{j}$, then we have $\mathbf{u} + \mathbf{v} = (a + c)\mathbf{i} + (b + d)\mathbf{j}$ and $\mathbf{u} \cdot \mathbf{v} = ac\mathbf{i}^2 + ad\mathbf{i} \cdot \mathbf{j} + bc\mathbf{j} \cdot \mathbf{i} + bd\mathbf{j}^2 = ac + bd$.

We now give an application of this combined approach.

Problem 1933.1.

Let a, b, c and d be real numbers such that $a^2 + b^2 = c^2 + d^2 = 1$ and $ac + bd = 0$. Determine the value of $ab + cd$.

Third Solution: Let $\mathbf{i}$ and $\mathbf{j}$ be two perpendicular unit vectors. Let $\mathbf{u} = a\mathbf{i} + b\mathbf{j}$ and $\mathbf{v} = c\mathbf{i} + d\mathbf{j}$. Note that we have $|\mathbf{u}|^2 = a^2 + b^2 = 1$, $|\mathbf{v}|^2 = c^2 + d^2 = 1$ and $\mathbf{u} \cdot \mathbf{v} = ac + bd = 0$. Hence $\mathbf{u}$ and $\mathbf{v}$ are also perpendicular unit vectors. Expressing $\mathbf{i}$ and $\mathbf{j}$ in terms of $\mathbf{u}$ and $\mathbf{v}$, we obtain $\mathbf{i} = (\mathbf{u} \cdot \mathbf{i})\mathbf{u} + (\mathbf{v} \cdot \mathbf{i})\mathbf{v} = a\mathbf{u} + c\mathbf{v}$ and $\mathbf{j} = (\mathbf{u} \cdot \mathbf{j})\mathbf{u} + (\mathbf{v} \cdot \mathbf{j})\mathbf{v} = b\mathbf{u} + d\mathbf{v}$. Now $ab + cd = \mathbf{i} \cdot \mathbf{j} = 0$ since $\mathbf{i}$ and $\mathbf{j}$ are perpendicular.

We can now give an alternative proof of Cauchy's Inequality in two variables. Let $\mathbf{u} = a\mathbf{i} + b\mathbf{j}$ and $\mathbf{v} = c\mathbf{i} + d\mathbf{j}$. Let θ be the angle between them. Then

$$ac + bd = \mathbf{u}\mathbf{v} = |\mathbf{u}||\mathbf{v}| \cos \theta \leq |\mathbf{u}||\mathbf{v}| = \sqrt{a^2 + b^2}\sqrt{c^2 + d^2}$$

or $(ac + bd)^2 \leq (a^2 + b^2)(c^2 + d^2)$.

The quadratic equation $x^2 + 1 = 0$ does not have solutions in real numbers. Nevertheless, it is useful to give it two solutions. We call one of them i. Then $i^2 = -1$. The other solution is obviously $-i$ since $(-i)^2 = i^2 = -1$.

A number of the form $a + bi$ where a and b are real numbers is called a **complex number**. In particular, i is a complex number, and all real numbers are complex numbers too.

The set of complex number is not ordered. In particular, i is neither positive nor negative. If we assume that $i > 0$, multiplying both sides by i yields $-1 = i^2 > 0$, which is a contradiction. Similarly, the assumption that $i < 0$ leads to the same contradiction, since the inequality must be reversed when we multiply both sides by the same negative number.

We can add and multiply complex numbers just like real numbers. We have

$$(a + bi) + (c + di) = (a + c) + (b + d)i$$

and

$$(a + bi)(c + di) = (ac - bd) + (ad + bc)i.$$

Note that the addition of complex numbers is just like vector addition.

A complex number $a + bi$ may be represented by the point (a, b) in the coordinate plane. The **modulus** of the complex number $a + bi$, the analog of the length of a vector, is defined as $|a+bi| = \sqrt{a^2 + b^2}$. The **argument** of the complex number $a + bi$ is defined as the angle between the x-axis and the segment joining $(0,0)$ to (a, b). It is taken to lie in the interval $[0°, 360°)$.

Let the modulus of $a + bi$ be r and its argument be θ. Then $a = r\cos\theta$ and $b = r\sin\theta$. Let the modulus of $c + di$ be s and its argument be ϕ. Then $c = s\cos\phi$ and $d = s\sin\phi$. It follows that

$$(a + bi) + (c + di)$$
$$= r(\cos\theta + i\sin\theta)s(\cos\phi + i\sin\phi)$$
$$= rs\big[(\cos\theta\cos\phi - \sin\theta\sin\phi) + i(\sin\theta\cos\phi + \cos\theta\sin\phi)\big]$$
$$= rs\big[\cos(\theta + \phi) + i\sin(\theta + \phi)\big].$$

In other words, when multiplying two complex numbers, we multiply their moduli and add their arguments. From this, it is easy to prove by induction that $\big[r(\cos\theta + i\sin\theta)\big]^n = r^n(\cos n\theta + i\sin n\theta)$ for any positive integer n. This is known as **de Moivre's Formula**.

When a complex number $a + bi$ is multiplied by another whose modulus is 1, the product is the point obtained by rotating $a + bi$ about the origin through an angle equal to the argument of the multiplier. Thus complex numbers are a very useful tool in handling geometric problems involving rotations.

A useful multiplier is the **primitive nth root of unity** with modulus 1 and argument $\frac{360°}{n}$. In particular, the primitive fourth root of unity is the imaginary unit $i = 0 + 1i$ represented by the point $(0, 1)$. It has modulus 1 and argument $90°$. Hence the point that represents the product ui is obtained by rotating the point u through $90°$. We give an application of this approach.

Problem 1933.1.

Let a, b, c and d be real numbers such that $a^2 + b^2 = c^2 + d^2 = 1$ and $ac + bd = 0$. Determine the value of $ab + cd$.

Fourth Solution: Let $\mathbf{i}$ and $\mathbf{j}$ be two perpendicular unit vectors. Let $\mathbf{u} = a\mathbf{i} + b\mathbf{j}$ and $\mathbf{v} = c\mathbf{i} + d\mathbf{j}$. Note that we have $|\mathbf{u}|^2 = a^2 + b^2 = 1$, $|\mathbf{v}|^2 = c^2 + d^2 = 1$ and $\mathbf{u} \cdot \mathbf{v} = ac + bd = 0$. Hence $\mathbf{u}$ and $\mathbf{v}$ are also perpendicular unit vectors. It follows that either $c + di = (a + bi)i$ or

$a + bi = (c + di)i$. In other words, either $c = -b$ and $d = a$ or $c = b$ and $d = -a$. Both lead to $ab + cd = ab - ab = 0$.

6.1.5 Solid Geometry

The geometry of the third dimension has completely vanished from the North American high school curriculum. The subject has also lost much of its earlier prominence even in mathematics competitions. Nevertheless, we do live in a three-dimensional world, and everyone should have some knowledge of the basic incidence properties of points, lines and planes. Surprisingly, this allows us to solve a number of problems in solid geometry, since much of the work is done by restricting our attention to specific plane sections.

In the plane, two points determine a unique line. Through a point not on a line ℓ, there exists a unique line parallel to ℓ. In space, three *noncollinear* points determine a unique plane. Through a point not on a plane Π, there exists a unique plane parallel to Π.

In the plane, two lines either intersect at a unique point or are parallel. In space, it is possible for two nonparallel lines not to intersect. Such a pair of lines are called **skew lines**. A practical example consists of a straight bridge over a straight river. They are clearly not parallel, but they most certainly do not intersect. In order for two lines in space to intersect, they must be *coplanar* and not be parallel.

A line ℓ not on a plane Π may be parallel to Π. If not, it will intersect Π at a unique point P. The line ℓ is said to be perpendicular to Π if it makes a right angle with every line on Π through P. Actually, it is sufficient if it does so with two such lines. Through any point, there exists a unique line perpendicular to a given plane. Through any point, there exists a unique plane perpendicular to a given line.

Two planes either intersect along a unique line or are parallel. Suppose they intersect along a line ℓ. Take any point P on ℓ and consider the plane through P perpendicular to ℓ. This plane intersects each of the two given planes in a line. The angle formed by these two lines is called the **dihedral angle** of the two planes. If it is a right angle, then the two planes are perpendicular to each other.

In the plane, the points equidistant from two given points form the perpendicular bisector of the segment joining the two points. In space, the points equidistant from two given points form a plane through the midpoint of the segment joining the two points and perpendicular to that segment.

In the plane, the points at a fixed distance from a given point form a circle with that point as center and that distance as radius. A line that is not disjoint from a given circle either touches it at one point or intersects it in two points. In space, the points at a fixed distance from a given point form a **sphere** with that point as center and that distance as radius. A plane that is not disjoint from a sphere either touches it at one point or intersects it along a circle.

6.2 Solutions

Problem 1932.2.

In triangle ABC, $AB \neq AC$. Let AF, AP and AT be the median, interior angle bisector and altitude from vertex A, with F, P and T on BC or its extension.

(a) Prove that P always lies between F and T.

(b) Prove that $\angle FAP < \angle PAT$ if ABC is an acute triangle.

Solution:

(a) In Figure 6.2.1, let D be the midpoint of the arc BC not containing A on the circumcircle of triangle ABC. Then DF is perpendicular

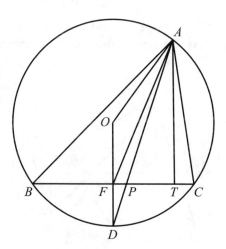

FIGURE 6.2.1

to BC and is therefore parallel to AT. Since the arcs BD and CD are equal, AD is the bisector of $\angle CAB$, so that P lies between A and D. It follows that P lies between the parallel lines DF and AT. Since it also lies on the line FT, it must lie between F and T.

(b) Let O be the circumcenter of triangle ABC. Since the triangle is acute, O is inside. In particular, F lies between O and D. It follows that $\angle FAP < \angle OAD = \angle ODA = \angle PAT$.

Problem 1933.3.

The circles k_1 and k_2 are tangent at the point P. A line is drawn through P, cutting k_1 at A_1 and k_2 at A_2. A second line is drawn through P, cutting k_1 at B_1 and k_2 at B_2. Prove that the triangles PA_1B_1 and PA_2B_2 are similar.

First Solution: In Figure 6.2.2, let O_1 be the center of k_1 and O_2 be that of k_2. Suppose the circles are tangent externally. Then

$$\angle O_1 A_1 P = \angle O_1 P A_1 = \angle O_2 P A_2 = \angle O_2 A_2 P,$$

so that triangles $O_1 A_1 P$ and $O_2 A_2 P$ are similar by the AA Theorem. It follows that $\frac{PA_1}{PA_2} = \frac{PO_1}{PO_2}$. Since triangles $O_1 B_1 P$ and $O_2 B_2 P$ are also similar,

$$\frac{PB_1}{PB_2} = \frac{PO_1}{PO_2} = \frac{PA_1}{PA_2}.$$

Moreover, $\angle A_1 P B_1 = \angle A_2 P B_2$, and triangles $PA_1 B_1$ and $PA_2 B_2$ are similar by the sAs Theorem. If the circles are tangent internally, the argument is exactly the same.

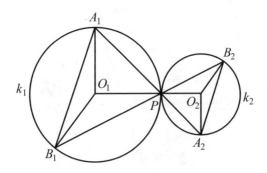

FIGURE 6.2.2

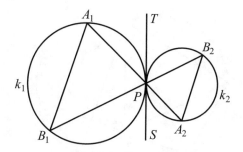

FIGURE 6.2.3

Second Solution: Let ST be the common tangent of k_1 and k_2 at P. Suppose the circles are tangent externally, as shown in Figure 6.2.3. Now

$$\angle PA_1B_1 = \angle SPB_1 = \angle TPB_2 = \angle PA_2B_2$$

by the Tangent-Angle Theorem. Together with

$$\angle A_1PB_1 = \angle A_2PB_2,$$

triangles A_1B_1P and A_2B_2P are similar by the AA Theorem. If the circles are tangent internally instead, we have

$$\angle PA_1B_1 = \angle SPB_1 = \angle SPB_2 = \angle PA_2B_2,$$

but the rest of the argument is the same.

Problem 1939.3.

ABC is an acute triangle. Three semicircles are constructed outwardly on the sides BC, CA and AB respectively. Construct points A', B' and C' on these semicircles respectively so that $AB' = AC'$, $BC' = BA'$ and $CA' = CB'$.

First Solution: Since ABC is acute, each altitude will intersect the semicircle on the opposite side as diameter. Call these points of intersection A', B' and C', as shown in Figure 6.2.4. Since $\angle BFC = 90° = \angle BEC$, $BCEF$ is a cyclic quadrilateral by the converse of the Circle-Angle Theorem. By the Intersecting Secants Theroem, $AF \cdot AB = AE \cdot AC$. By the Semicircle-Angle Theorem, $\angle AC'B = 90° = \angle AFC$. Since $\angle BAC' = \angle C'AF$, triangles BAC' and $C'AF$ are similar by the AA Theorem. It follows that $\frac{BA}{AC'} = \frac{C'A}{AF}$, so that $C'A^2 = AF \cdot AB$. Similarly, $B'A^2 = AE \cdot AC$. It follows that $C'A = B'A$. The other two equalities can be established in the same way.

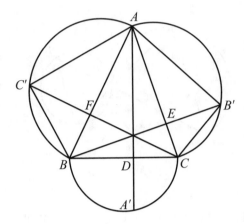

FIGURE 6.2.4

Second Solution: Since ABC is acute, the altitude from A cuts the semicircle on BC as diameter at a point A'. Draw a circle through A' with center B, cutting the semicircle on AB as diameter at a point C'. Draw another circle through A' with center C, cutting the semicircle on CA as diameter at a point B'. The common chord of these two circles is their radical axis. Since it is perpendicular to BC, it passes through A, as illustrated in Figure 6.2.5. By the Semicircle-Angle Theorem, $\angle AC'B = 90° = \angle AB'C$. Hence AC' and AB' are tangents to the respective circles

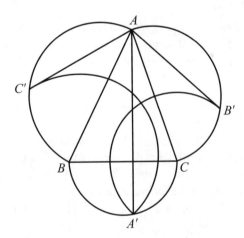

FIGURE 6.2.5

by the converse of the Tangent-Radius Theorem. Since A lies on the radical axis of the two circles, we have $AC' = AB'$. The other two equalities can be established in the same way.

Remark: For another approach to this problem, see page 133 in this subsection.

Problem 1941.3.

The hexagon $ABCDEF$ is inscribed in a circle. The sides AB, CD and EF are all equal in length to the radius. Prove that the midpoints of the other three sides determine an equilateral triangle.

First Solution: Let P, Q, R, U, V, W and X be the midpoints of BC, DE, FA, OA, OB, OC and OD respectively. By the Midpoint Theorem, $UV = \frac{1}{2}AB = \frac{1}{2}OB = PW$. Similarly, $PV = WX$. Since UV is parallel to AB and PW to OB, the angle between UV and PW is 60°. Similarly, the angle between PV and WX is also 60°. Thus in Figure 6.2.6, rotating $\angle UVP$ clockwise about V through an angle of 120° takes its arms UV and VP respectively to lie in the directions of the arms PW and WX of $\angle PWX$. Hence the angles are equal. It follows that UVP and PWX are congruent triangles, with $UP = PX$. This rotation also takes

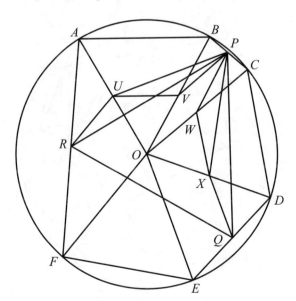

FIGURE 6.2.6

UP in the direction to PX, and we have $\angle UPX = 180° - 120° = 60°$. Now $RU = \frac{1}{2}OF = \frac{1}{2}OE = QX$. Since RU is parallel to OF and QX to OE, the angle between RU and QX is $60°$. It follows that PUR and PXQ are congruent triangles, with $PR = PQ$ and $\angle RPQ = 60°$. Hence PQR is an equilateral triangle.

Second Solution: In Figure 6.2.7, let M, P, N, Q, L and R be the midpoints of AB, BC, CD, DE, EF and FA respectively. Since AB is equal to the radius of the circle, the arc AB subtends an angle of $60°$ at the center of the circle. By Thales' Theorem, $\angle ACB = 30°$. Applying the Midpoint Theorem to triangle ABC, $MP = \frac{1}{2}AC$ and $\angle MPB = 30°$. Similarly, $\angle NPC = 30°$ and $NP = \frac{1}{2}BD$. Hence $\angle MPN = 120°$. Since $AB = CD$, we have $AC = BD$ by symmetry, so that $MP = NP$. Similarly, $NQ = LQ$ and $LR = MR$. Let the circular arc from M to N with center P intersect the circular arc from N to L with center Q at the point K. By Thales' Theorem and the Opposite Angles Theorem, $\angle MKN = 180° - \frac{1}{2}\angle MPN = 120°$. Similarly, we have $\angle NKL = 120°$. Hence $\angle LKM = 360° - \angle MKN - \angle NKL = 120°$ also. We can prove that $\angle LRM = 120°$ in exactly the same way that we proved $\angle MPN = 120°$. It follows that the circular arc from L to

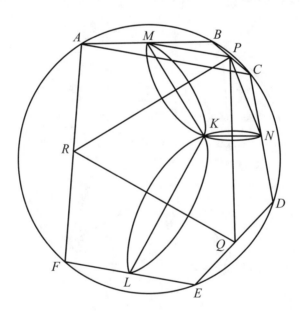

FIGURE 6.2.7

M with center R also passes through K. Now KL, KM and KN are perpendicular to the sides of triangle PQR since they are common chords of the arcs. Now $\angle RPQ = 180° - \angle MKN = 60°$ by the Opposite Angles Theorem. Similarly, $\angle PQR = \angle QRP = 60°$, so that PQR is an equilateral triangle.

Third Solution: Let P, Q and R be the midpoints of BC, DE and FA respectively. Suppose PQR is an equilateral triangle for a particular hexagon $ABCDEF$ with AB, CD and EF equal to the radius of the circle. In Figure 6.2.8, rotate the equilateral triangle OEF about O to $OE'F'$. Let Q' and R' be the midpoints of DE' and $F'A$ respectively. By the Midpoint Theorem, QQ' is parallel to EE' and equal to half its length, while RR' is parallel to FF' and equal to half its length. Now a $60°$ rotation about O sends EE' to FF'. Hence a $60°$ rotation about some point should send QQ' to RR'. This point can only be P since $PQ = PR$ and $\angle RPQ = 60°$. It follows that $PQ' = PR'$ and $\angle R'PQ' = 60°$, so that $PQ'R'$ is also an equilateral triangle. Now PQR is certainly equilateral if $ABCDEF$ is a regular hexagon. We can rotate triangles OCD and OEF successively so that the given hexagon is obtained. Since the rotations do not affect our conclusion, PQR is always an equilateral triangle.

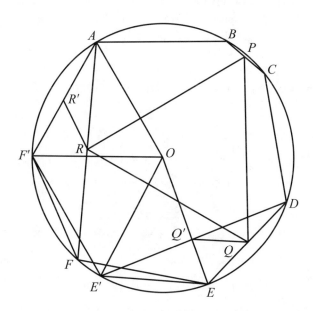

FIGURE 6.2.8

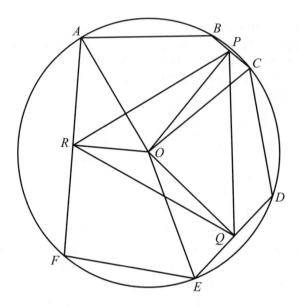

FIGURE 6.2.9

Fourth Solution: In Figure 6.2.9, let P, Q and R be the midpoints of BC, DE and FA respectively. We may take the radius of the circle to be 1. Let $\angle COP = \alpha$, $\angle EOQ = \beta$ and $\angle AOR = \gamma$. The chords BC, DE and FA respectively subtend angles of 2α, 2β and 2γ at the center O. Hence the 360° around O consists of $2(\alpha + \beta + \gamma)$ and three 60° angles. It follows that $\alpha + \beta + \gamma = 90°$, so that $\alpha + 60° + \beta = 150° - \gamma$. From the Law of Cosines,

$$PQ^2 = OP^2 + OQ^2 - 2OP \cdot OQ \cos POQ$$

$$= \cos^2 \alpha + \cos^2 \beta - 2 \cos \alpha \cos \beta \cos(150° - \gamma)$$

$$= \cos^2 \alpha + \cos^2 \beta + \sqrt{3} \cos \alpha \cos \beta \cos \gamma - \cos \alpha \cos \beta \sin \gamma$$

$$= \cos^2 \alpha + \cos^2 \beta + \cos^2 \gamma - 1$$

$$\quad + \sqrt{3} \cos \alpha \cos \beta \cos \gamma + \sin \gamma \big[\cos(\alpha + \beta) - \cos \alpha \cos \beta \big]$$

$$= \cos^2 \alpha + \cos^2 \beta + \cos^2 \gamma - 1$$

$$\quad + \sqrt{3} \cos \alpha \cos \beta \cos \gamma - \sin \alpha \sin \beta \sin \gamma.$$

The last expression is symmetric with respect to α, β and γ. Hence $PQ = QR = RP$ and PQR is an equilateral triangle.

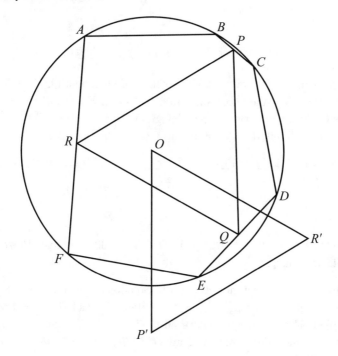

FIGURE 6.2.10

Fifth Solution: Let O be the origin in the complex plane. Let the point A be represented by the complex number a, and so on. We have $p = \frac{1}{2}(b+c)$, $q = \frac{1}{2}(d+e)$ and $r = \frac{1}{2}(f+a)$. In Figure 6.2.10, let P' be the point such that $OPQP'$ is a parallelogram. Then P' is represented by $q - p = \frac{1}{2}(d+e-b-c)$. Let ω be a primitive sixth root of unity. Then $0 = \omega^3 + 1 = (\omega+1)(\omega^2 - \omega + 1)$. Since $\omega \neq 1$, we have $\omega = \omega^2 + 1$. Note that $\omega d = c$, $\omega b = a$, $\omega^2 e = -f$ and $\omega^2 c = -d$. Routine calculation yields $\omega(q-p) = \frac{1}{2}(d+e-f-a) = q-r$. Now $q-r$ represents the point R' where $ORQR'$ is a parallelogram. Since multiplication by ω amounts to a $60°$ rotation about O, we have $P'O = OR'$ and $\angle P'OR' = 60°$. It follows that $PQ = QR$ and $\angle PQR = \angle P'OR' = 60°$, so that PQR is an equilateral triangle.

Sixth Solution: We use a vectorial approach. Since P is the midpoint of BC, we have $\overline{OP} = \frac{1}{2}(\overline{OB} + \overline{OC})$. Similarly, we have

$$\overline{OQ} = \frac{1}{2}(\overline{OD} + \overline{OE}) \quad \text{and} \quad \overline{OR} = \frac{1}{2}(\overline{OF} + \overline{OA}).$$

Hence

$$\overline{PQ} = \overline{OQ} - \overline{OP} = \frac{1}{2}(\overline{OD} + \overline{OE} - \overline{OB} - \overline{OC}).$$

Consider now a counterclockwise rotation of $60°$ about O. Note that $\overline{OD}$ turns into $\overline{OC}$ while $\overline{OB}$ turns into $\overline{OA}$. Now $\overline{OE}$ turns into $\overline{OE'}$ where E' is the fourth vertex of the parallelogram $OFEE'$. Hence $\overline{OE'} = \overline{FE} = \overline{OE} - \overline{OF}$. Similarly, $\overline{OC}$ turns into $\overline{OC} - \overline{OD}$. We have

$$\frac{1}{2}(\overline{OC} + \overline{OE} - \overline{OF} - \overline{OA} - \overline{OC} + \overline{OD}) = \frac{1}{2}(\overline{OD} + \overline{OE} - \overline{OF} - \overline{OA})$$

$$= \overline{OQ} - \overline{OR}$$

$$= \overline{RQ}.$$

It follows that $PQ = RQ$ and $\angle PQR = 60°$, so that PQR is an equilateral triangle.

Remark: The First and the Fifth Solutions appeared in the paper "The Asymmetric Propellers" by Leon Bankoff, Paul Erdős and Murray Klamkin, *Mathematics Magazine* **46** (1973) 270–272. See also the paper with the same title by Martin Gardner, *College Mathematics Journal* **30** (1999) 18–22.

Problem 1934.2.

Which polygon inscribed in a given circle has the property that the sum of the squares of the lengths of its sides is maximum?

First Solution: Consider any n-gon inscribed in the given circle, where $n \geq 4$. Then the sum of its angles is $(n - 2)180°$. By the Mean Value Principle, there is at least one angle that is no less than $\frac{n-2}{n}180° \geq 90°$. We may assume that it is $\angle ABC$, and we reduce the n-gon to an $(n - 1)$-gon by drawing the segment AC, cutting off triangle ABC. By Pythagoras' Inequality, $AC^2 \geq AB^2 + BC^2$. Since we want the sum of the squares of the side-lengths to be as large as possible, we can ignore the n-gon and consider the resulting $(n - 1)$-gon instead. It follows that we only have to make comparisons among the triangles.

In Figure 6.2.11, let QR be a chord of a circle that is not a diameter. Let MN be the diameter passing through the midpoint S of QR, with M on the same side as the center O of the circle. Let P be a point moving along the circle from N towards M. In triangle SOP, OS and OP are constant while $\angle SOP$ is increasing. By the sAs Inequality, SP is also increasing. Applying the Median Theorem to triangle PQR, we

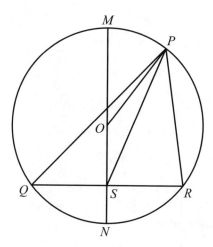

FIGURE 6.2.11

have $PQ^2 + PR^2 = 2QS^2 + 2SP^2$. Since QS is constant, $PQ^2 + PR^2$ is increasing.

Let ABC be a nonequilateral triangle inscribed in the given circle, with $AB \geq BC \geq CA$. Then BC is not a diameter and the arc AC is less than one-third of the circle. Move A away from C towards B until either the arc CA or the arc AB is exactly one-third of the circle, whichever situation arises first. Then A would not have passed over the midpoint of the major arc BC, so that the sum of the squares of the side-lengths of triangle ABC has increased by the result in the preceding paragraph. If the new ABC is still not equilateral, then one of AB and BC is less than one-third of the circle. By fixing AC, we can move B so that ABC becomes an equilateral triangle while the sum of the squares of its side-lengths increases again. It follows that of all polygons inscribed in a given circle, the equilateral triangle has the largest sum of the squares of its side-lengths.

Second Solution: As in the first solution, we can restrict our attention to triangles. Let BC, CA and AB subtend angles 2α, 2β and 2γ at the center of the circle in which ABC is inscribed. By the Law of Sines,

$$BC^2 + CA^2 + AB^2 = 4R^2(\sin^2 \alpha + \sin^2 \beta + \sin^2 \gamma),$$

where R is the radius of the circle. Since $4R^2$ is constant, we only have to maximize the second factor. Using trigonometric identities, we can rewrite it as:

$$\sin^2 \alpha + \sin^2 \beta + \sin^2 \gamma$$

$$= 1 - \cos^2 \alpha + \frac{1}{2}(1 - \cos 2\beta) + \frac{1}{2}(1 - \cos 2\gamma)$$

$$= 2 - \cos^2 \alpha - \cos(\beta + \gamma)\cos(\beta - \gamma)$$

$$= 2 - \cos^2 \alpha + \cos \alpha \cos(\beta - \gamma)$$

$$= 2 - \left[\cos \alpha - \frac{1}{2}\cos(\beta - \gamma)\right]^2 + \frac{1}{4}\cos^2(\beta - \gamma).$$

The last expression is maximum if and only if $\cos \alpha = \frac{1}{2}\cos(\beta - \gamma)$ and $\cos(\beta - \gamma) = 1$. This is equivalent to $\beta = \gamma$ and $\alpha = 60°$. In other words, ABC is an equilateral triangle.

Problem 1941.2.

Prove that if all four vertices of a parallelogram are lattice points and there are some other lattice points in or on the parallelogram, then its area exceeds 1.

First Solution: By the Coordinates-Area Formula, the area of a lattice triangle is $\frac{1}{2}$ times an integer. Since it is positive, it is at least $\frac{1}{2}$. In the given parallelogram, join the fifth lattice point to all four vertices. This divides the lattice parallelogram into at least three lattice triangles, each of which has area at least $\frac{1}{2}$. Hence the area of the lattice parallelogram exceeds 1.

Second Solution: By Pick's Theorem, the area of a lattice polygon is given by $I + \frac{1}{2}O - 1$, where I stands for the number of lattice points in its interior and O stands for the number of lattice points on its boundary. The given parallelogram contains at least five lattice points. Even if they are all on the boundary and there are no others, the area of the lattice parallelogram will be $\frac{1}{2}(5) - 1 = \frac{3}{2}$. If some of them are in the interior or if there are other lattice points, the area will only get larger. Hence the area of the lattice parallelogram exceeds 1.

Problem 1932.3.

Let α, β and γ be the interior angles of an acute triangle. Prove that if $\alpha < \beta < \gamma$, then $\sin 2\alpha > \sin 2\beta > \sin 2\gamma$.

First Solution: Let ABC be the triangle with $\angle CAB = \alpha$, $\angle ABC = \beta$ and $\angle BCA = \gamma$. Reflect B across CA to B' and reflect C across AB

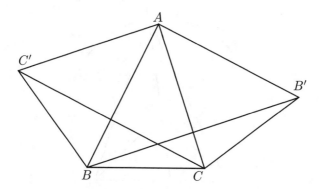

FIGURE 6.2.12

to C', as shown in Figure 6.2.12. Since ABC is acute, both $AC'BC$ and $ABCB'$ are convex quadrilaterals. They have equal areas since each is twice that of ABC. Since $\beta < \gamma$, we have $AB > AC$ by the Angle-Side Inequality. By the Sine-Area Formula,

$$[BAB'] = \frac{1}{2}AB^2 \sin 2\alpha > \frac{1}{2}AC^2 \sin 2\alpha = [CAC'].$$

Note that $[BAB'] + [AC'B] > [CAC'] + [ACB']$. This is because we have $[ABC'] = [ABC] = [ACB']$. Hence $[AC'BB'] > [AC'CB']$. By the Sine-Area Formula,

$$\frac{1}{2}BC^2 \sin 2\gamma = [BCB'] < [CBC'] = \frac{1}{2}BC^2 \sin 2\beta,$$

so that $\sin 2\gamma < \sin 2\beta$. Similarly, $\sin 2\beta < \sin 2\alpha$.

Second Solution: Since γ is an acute angle, $\alpha + \beta > 90°$ so that $\alpha > 90° - \beta$. Since $\alpha < \beta$, $2\beta > 90°$. Hence $180° - 2\beta < 2\alpha < 2\beta$. Now $\sin(180° - 2\beta) = \sin 2\beta$, and the sine of any angle between 2β and $180° - 2\beta$ has a greater value. It follows that $\sin 2\alpha > \sin 2\beta$. On the other hand, we have $90° < 2\beta < 2\gamma < 180°$. It follows that $\sin 2\beta > \sin 2\gamma$ since sine is a decreasing function on the interval $[90°, 180°]$.

Third Solution: Let ABC be the triangle with $\angle CAB = \alpha$, $\angle ABC = \beta$ and $\angle BCA = \gamma$. Let O be its circumcenter. Since ABC is acute, the extension of AO cut BC at an interior point D, as shown in Figure 6.2.13. Since $\beta < \gamma$, we have $AB > AC$ so that $BD < CD$. The triangles BOD and COD have the same altitude from O, so that their areas are proportional to their bases BD and CD. It follows from the Sine-Area

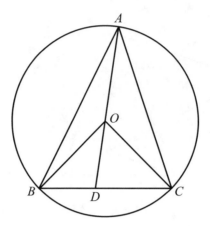

FIGURE 6.2.13

Formula that

$$\frac{1}{2}OB \cdot OD \sin(180° - 2\gamma) = [BOD] < [COD] = \frac{1}{2}OC \cdot OD \sin(180° - 2\beta).$$

Hence $\sin 2\gamma < \sin 2\beta$, and similarly, $\sin 2\beta < \sin 2\alpha$.

Fourth Solution: We have $\sin 2\beta - \sin 2\alpha = 2\cos(\beta + \alpha)\sin(\beta - \alpha) = -2\cos\gamma \sin(\beta - \alpha)$. Since both γ and $\beta - \alpha$ are acute angles, $\cos\gamma > 0$ and $\sin(\beta - \alpha) > 0$. It follows that $\sin 2\alpha > \sin 2\beta$. Similarly, $\sin 2\beta > \sin 2\gamma$.

Fifth Solution: Since α is an acute angle, $180° - 2\alpha$ is a positive angle. Similarly, so are $180° - 2\beta$ and $180° - 2\gamma$. Since they add up to $180°$, there exists a triangle PQR such that $\angle RPQ = 180° - 2\alpha$, $\angle PQR = 180° - 2\beta$ and $\angle QRP = 180° - 2\gamma$. We have $180° - 2\alpha > 180° - 2\beta > 180° - 2\gamma$ since $\alpha < \beta < \gamma$. By the Angle-Side Inequality, $QR > RP > PQ$. Applying the Law of Sines to PQR, we have

$$\frac{\sin(180° - 2\alpha)}{QR} = \frac{\sin(180° - 2\beta)}{RP} = \frac{\sin(180° - 2\gamma)}{PQ}.$$

It follows that $\sin 2\alpha > \sin 2\beta > \sin 2\gamma$.

Problem 1938.3.

Prove that for any acute triangle, there is a point in space such that every line segment from a vertex of the triangle to a point on the line joining the other two vertices subtends a right angle at this point.

First Solution: First, note that the sides of triangle ABC are also cevians. Hence the point O we seek satisfies $\angle AOB = \angle BOC = \angle COA = 90°$. This means that OC is perpendicular to the plane OAB. If CP is a cevian of ABC, then P lies on AB and therefore in the plane OAB. It follows that $\angle COP = 90°$. Thus every cevian through C, and by symmetry every cevian of ABC, subtends a right angle at O.

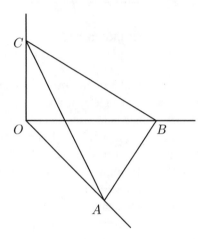

FIGURE 6.2.14

Let $BC = a$, $CA = b$ and $AB = c$. Since ABC is acute, Pythagoras' Inequality yields $b^2 + c^2 > a^2$, $c^2 + a^2 > b^2$ and $a^2 + b^2 > c^2$. Take three mutually perpendicular lines meeting at a point O. Take one point on each line so that $OA = x$, $OB = y$ and $OC = z$, as shown in Figure 6.2.14. We claim that x, y and z can be chosen so that $BC = a$, $CA = b$ and $AB = c$. By Pythagoras' Theorem, x, y and z must satisfy $y^2 + z^2 = a^2$, $z^2 + x^2 = b^2$ and $x^2 + y^2 = c^2$. Solving this system of simultaneous equations, we have

$$x = \sqrt{\frac{b^2 + c^2 - a^2}{2}}, y = \sqrt{\frac{c^2 + a^2 - b^2}{2}} \text{ and } z = \sqrt{\frac{a^2 + b^2 - c^2}{2}}.$$

Second Solution: As in the First Solution, we seek a point O in space that satisfies $\angle AOB = \angle BOC = \angle COA = 90°$. Now a point P that satisfies $\angle APB = 90°$ lies on the sphere with AB as diameter. If the spheres with the sides of triangle ABC as diameters have a common point, this may be taken as O.

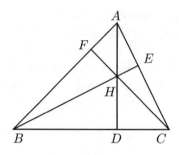

FIGURE 6.2.15

In Figure 6.2.15, let AD and BE be two of the altitudes of ABC, meeting at the orthocenter H of the triangle. Consider for now the spheres with AB and AC as diameters. Since both are symmetrical with respect to the plane ABC, they intersect along a circle that lies in a plane perpendicular to ABC. Now A lies on both spheres. Since $\angle ADB = 90° = \angle ADC$, D also lies on both spheres. It follows that the circle of intersection has AD as diameter since it is symmetrical with respect to the plane ABC. Similarly, the spheres with AB and BC as diameters intersect along a circle with BE as diameter. The three spheres have a common point if and only if so do these two circles. Note that both circles lie on the sphere with AB as diameter. Since ABC is acute, H is inside the triangle. Hence the line through H perpendicular to ABC intersects this sphere at two points which must lie on both circles. Either of these points may be taken as O.

Remark: This problem is closely related to **Problem 1939.3**. In the second solution to that problem, if we fold the triangles ABC', BCA' and CAB' along the sides of triangle ABC until AC' coincides with AB' and BA' coincides with BC', then CA' will coincide with CB', the points A', B' and C' merging to form the point O in the second solution to this problem. On the other hand, if we cut open the tetrahedron $OABC$ along the sides OA, OB and OC, and fold the lateral faces until they are coplanar with the base, we have the desired conclusion of the other problem.

Problem 1937.2.

Two circles in space are said to be tangent to each other if they have a common tangent at the same point of tangency. Assume that there are three circles in space that are mutually tangent at three distinct points. Prove that they either all lie in a plane or all lie on a sphere.

Solution: Define the *axis* of a circle as the line passing through its center and perpendicular to its plane. Consider first two tangent circles. Their axes both lie on the plane passing through the point of tangency and perpendicular to the common tangent, so that they are not skew lines. If they are parallel to each other, then the two circles lie on the same plane. If they are not parallel, then they intersect at some point. This point is equidistant to every point on each circle. Since the point of tangency is common to both circles, they lie on the same sphere.

Consider now three circles. If the first two lie in a plane and the last two also lie in a plane, then all three lie on the same plane. Suppose the first two circles lie on a sphere S. If the last two circles lie in a plane or a sphere other than S, then this plane or sphere intersects S precisely along the second circle. Hence the point of tangency of the first and the third circles must lie on the second one. Since the second circle has exactly one point in common with each of the other two circles, this point is the common point of tangency for all three, which is contrary to the hypothesis that the three points of tangency are distinct.

Theorem Index

Term Index

Problem Index

Year	#1		#2		#3	
1929	CB	26	AB	60	G1	97
1930	CB	25	CB	24	G1	88
1931	NT	43	NT	43	G1	86
1932	NT	45	G2	119	G2	130
1933	AB	55	CB	24	G2	120
1934	NT	45	G2	128	CB	30
1935	AB	58	CB	29	CB	28
1936	AB	56	G1	92	NT	46
1937	AB	58	G2	134	G1	87
1938	NT	42	AB	58	G2	132
1939	AB	60	NT	44	G2	121
1940	CB	22	NT	43	G1	91
1941	AB	56	G2	130	G2	123
1942	G1	87	NT	42	G1	94
1943	CB	23	G1	89	AB	57

See page xvii on how to read this chart.